T0140223

Climate-Drought Resilience in Extreme Environments

V. Ratna Reddy • Y. V. Malla Reddy
M. Srinivasa Reddy

Climate-Drought Resilience in Extreme Environments

 Springer

V. Ratna Reddy
Livelihoods and Natural Resource
Management Institute
Hyderabad, India

Y. V. Malla Reddy
Accion Fraterna Ecology Centre
Anantapur, Andhra Pradesh, India

M. Srinivasa Reddy
Centre for Economic and Social Studies
Hyderabad, India

ISBN 978-3-030-45891-1   ISBN 978-3-030-45889-8   (eBook)
https://doi.org/10.1007/978-3-030-45889-8

This Springer imprint is published by the registered company Springer Nature Switzerland AG.
The registered company address is: Gewerbestrasse 11, 6330 Cham, Switzerland

Foreword

This book deals with the evolution of watershed development (WSD) in the state of Andhra Pradesh, a history that also tells us of its future relevance to India. Extremely well-researched and full of authentic data, all of it primary and from the field, this book will serve to open the eyes of those concerned with making policy relating to agriculture, water management, food and nutrition security, rural employment, migration and participative democratic self-governance. I say this for many reasons, including the fact that the question whether there was going to be a serious possibility of a veritable exodus from agriculture as an occupation itself was raised by Prime Minister Narasimha Rao in the mid-1990s. The seriousness of this would later manifest itself as a phenomenon of continual suicides by farmers, including in Andhra Pradesh, a situation taken note of by India's National Human Rights Commission that ordered an investigation into it with me as its Special Rapporteur. Therefore, a study like this one that situates watershed development in the context of issues that go far beyond its original conceptualisation in 1983 so as to include larger environmental and livelihood issues is one that is critical to the future of the lives of people who inhabit geographies such as Anantapur. The facts that relate to these lives are that only a third of India's agricultural land has assured irrigation, the rest being dependent on uncertain monsoons. Only half the area growing food grains in India has irrigation. Two thirds of India's irrigation comes from groundwater, which is currently depleting rapidly. India, therefore, is essentially a rainfed agricultural country. It is in the rainfed agriculture regions of India that its poverty mostly resides, with low Human Development Indices. Naturally enough, interstate and intrastate migrations in search of food and employment are endemic to us, manifesting many shades of bonded labour as well.

When COVID-19 struck, the country stood exposed to this. Hunger was its prime manifestation with its natural consequence.

This book therefore constitutes a response to a crisis that India faces in about half its total land mass. The message of this book emanates from the implementation of the Watershed Development Programme for over 30 years by Accion Fraterna Ecology Centre, Anantapur, in the full view of all stakeholders.

Midway between the time we became free in 1947 and now, the Watershed Management approach was born to confront this problem manifesting itself as routine droughts, prevent soil erosion and improve moisture availability, and make agriculture sustainable in these areas. This book authentically shows that such expectation was not realised, calling for drastically holistic changes to the approach to watershed development. And what it also credibly shows us is the way forward to achieving the objectives the Watershed Movement had set for itself, and also beyond. The significant message the authors give us is that WSD is as relevant as ever, still at the "centre", and a necessary condition as a participatory, integrated watershed programme involving equal participation of women and other vulnerable sections of the community for improving agriculture in these naturally disadvantaged regions. However, this is not sufficient by itself, especially in the context of climate change. Water and land are still inseparable Siamese twins but in dire requirement of a paradigm shift by way of the recognition that there are no perfect solutions to issues nature confronts us with, without involving human and social development interventions. In this context, the key concept that the authors have employed for assessing the innovative interventions we need, in pursuit of their objectives in writing this book, is aligning WSD with the five capitals that constitute the framework for assessing sustainable livelihoods and resilient impacts on households in drought-induced crises. These capitals are: Natural, Physical, Financial, Human and Social. This alignment should generate "Resilience", "Integration", "Convergence", the recognition that WSD is and would remain a "process in action" calling for internal changes within it, and then integrated with the livelihoods strategies. These experiments have already delivered certain positive results as in some areas yields have increased and dependence on moneylenders has declined. Yet others are in the testing stage, but there has been the heart-warming policy-influencing intimation, for example, that the Government of AP allocated Rs160 crores for scaling up one of the proven strategies, namely, Protective Irrigation.

Several important lessons emerge from this book. On the face of it, it may appear that we know some of them already but what gives those a special validity is the authenticity of those lessons drawn not from text books but from field-level work in conditions that are as exhausting as they are for those women, children and men who experience them as if it is a routine fate that cannot be conquered, for example, the concept of Capacity Building as the key to better implementation and livelihoods. That observation looks commonplace till the authors tell you:

"Development of human capital in a harsh environment like Ananthapuramu is critical for enhancing livelihoods and alleviating poverty. While, there are limits for strengthening the natural capital base, due to geographical and climatic constraints, human capital development has greater potential to expand and alleviate the communities from the dire situation. AF-EC strongly believes in this and has focused and invested in capacity building over the years. AF-EC has an overarching approach to capacity building. All their staff is locally recruited and most of them come from the programme villages and many of them have long experience in the organization and have grown within. This automatically ensures better understanding of the local context and also commitment and concern for improving the conditions.

Communities also feel comfortable to approach them any time for any kind of support, irrespective of whether there is an ongoing programme in the village or not".

"AF-EC has about 60 committed and experienced socio technical organisers (STOs) at grass root level. They are all educated rural youth from the district. Many of them are trained and well experienced. Their training and experience include WSC, CBO formation and capacity building, participatory planning and implementation, community organization, etc., Their technical training and experience include a variety of WSD skills like SMC, RWH, horticulture, sustainable agriculture (SA) including rain-fed agronomical practices, bio-diversity, crop-diversity, bio/non-pesticidal management (NPM), bio-fertilizers (organic farming/ INM: integrated nutrient management) like composting, liquid fertilizers, alternate livelihoods for women and youth, etc."

"Community level capacity building is also given equal importance. Communities are capacitated not only on WSD but also on crop practices, crop technologies and livelihood activities. Its approach is not limited to specific programme or intervention/cadre or section of the community. Appropriate training was provided for all the sections of the community, i.e., watershed committee and village organisation members, SHG and UG members, landless households, Sasya Mitra (friend of plants or trees) and MACS members (Tables 3.1 and 3.2). Trainings and exposure visits on watershed and farming activities included new agricultural practices, crop systems, input management, horticulture development, maintenance of watershed structures, etc. WSC members were trained on technical aspects of watershed interventions and monitoring system of works. VO members and other farmers were trained on PSI methods. SHG and UG members are provided training on savings and investments, managing loan repayment, etc. Labourers are provided with technical training on farm pond construction and farming. They include alternative livelihoods like garment making, auto driving, managing credit coopera-tives, etc." Or, "Managing moisture stress to save the crop is the main challenge in Ananthapuramu district. Apart from crop systems like contingency crops, AF-EC has been promoting technologies and techniques in order to address drought. The initiatives include: (i) timely sowing of crops under insufficient soil moisture condi-tion, (ii) providing protective irrigation during prolonged dry spells, and (iii) pro-moting farm implements for drought mitigation and drudgery reduction".

Next, there are other interventions that translate technology into equipment based on field trials and tests.

Finally comes the conclusion, which I think is the essence of this book: "Given the harsh environmental and natural conditions, the contribution of agriculture to the household income is limited despite the fact that majority of the population depends on agriculture. Dependence on agriculture is mainly due to the absence of viable livelihood opportunities rather than by choice. Agriculture is increasingly becoming unviable in general and more severely in these regions. While strengthening the natural resource base, especially water and soils, is necessary to improve the viability and share of agriculture income, it may not be sufficient to ameliorate poverty in the given natural constraints. A multi-pronged approach is required to bring these regions out of the dire conditions. Strengthening the natural resources base through watershed interventions is a medium- to long-term option, enhancing the resilience of the communities is a short- to medium-term option and increasing the share of agricultural income or reducing the dependence on agriculture are long-term solutions".

"The AF-EC technical initiatives fostered with appropriate institutional support emphasizes the importance of both short- and long-term options. Given the fragility of the region in terms of natural resources and the associated financial fragility of the household economy, AF-EC has realized early on that natural resource-based interventions like WSD is a necessary but not a sufficient condition for improving the economic conditions of the region (Reddy et al. 2004). The watershed institutions along with other institutions not only sustain the programme interventions but also support the communities in various other activities that strengthen and sustain their livelihoods. Its present approach is comprehensive and integrated, viz., WSD (resource base) + savings and credit based and livelihoods support institutions (enhancing financial capital) + technical and policy support (drought resilient extension support and new programme interventions). The combined approach appears to be more effective in realizing the programme impacts and livelihood benefits. This approach goes much beyond the now widely adopted watershed plus approach and could be termed as 'watershed plus + plus'- (WS++)." A part of this approach could be strengthened by what the authors have proposed: "... there are above 3000 small and medium water bodies spread across in Ananthapuramu District. All these water bodies need to be revived and converted into percolation tanks. These are common property resources of the villages and all households have the right to use the water. If these can be filled with *Tungabhadra* or *Hundri Niva Sujala Sravanthi* (HNSS) project water, all farmers can use it for protective irrigation through borewells, as percolation tanks recharge and stabilize groundwater. Communities feel that it can support much needed protective irrigation needs in a most efficient way. This is a necessary condition for the evolution and success of groundwater collectivisation interventions and institutions in these harsh environmental conditions"...

"In the absence of assured water availability to protect at least one crop in a year it would be difficult to sustain farming in the district. For, one drought can wipe out all savings and income. In this context, there is a need to create water grid exclusively for Anantapuramu, i.e., '**Anantapuramu water grid**'.

Finally, the strategic future: Watershed Development and Beyond, where other supportive livelihood strategies provide for the resilience at the personal, households and community levels.

The authors have emphasized that AF-EC has been effectively drawing additional resources through integration of other related programmes like MGNREGS and Andhra Pradesh Drought Adaption Programmes (APDAP). "While a number of states have been struggling to achieve convergence of programmes AF-EC could achieve it smoothly through its participatory approaches. In fact, AF-EC has demonstrated to the state government the effectiveness of such integration and convergence. Similarly, all the initiatives, be it water related, land related or livelihoods related, are developed and implemented in an interlinked and integrated manner rather than in isolation. The effectiveness of an integrated approach has been demonstrated well".

A crucial point the authors make is this: "Sustaining these initiatives is not easy in the present policy environment, as there is no price support for these crops at the policy level. AF-EC has been lobbying for subsidies for intercrops at par with those extended to groundnut, and for inclusion of millets and pulses in fair price shops (FPS), school mid-day meal programme (MDMP), integrated child development services (ICDS) and in hostels in government schools and colleges. Replacing rice with locally produced millets and pulses also reduces the food miles in transporting rice." This requires to be widely internalised by all of us interested in the nutrition security of the country. To this, I would further add that the Minimum Support Prices for millets and the nutritive cereals grown in the dry lands of the country should be double that of the Food Corporation of India's purchasing price of fine rice under levy and all the production of these should be mandatorily purchased by the Food Corporation of India. Distribution of these in the PDS, ICDS and MDM should be promoted by widespread health and nutrition education. India's dry land agriculture should be backed and promoted as part of India's Health Policy.

The authors are modest enough to acknowledge that while "the comprehensive approach of integrating various dimensions of livelihoods like capacity building, integrated resource management, technologies and institutions are making communities resilient to drought… it may be naive or too early to say that the communities are resilient enough to withstand more than one drought, the process is set and moving in the right direction." They have also acknowledged that "while AF-EC has been focusing on appropriate crop systems, techniques and technologies, integration of hydrogeology into watershed designing and planning needs more efforts. Our hydrogeology assessment has revealed a mismatch between required and actual RWH interventions."

Theirs is a call to continued, sustained action.

Thus, this book has rich and relevant material to offer by way of hope for India's rainfed agriculture. The authors deserve our gratitude for courageously setting out what the problems are and how by similar courage and conviction in our own ability and through the social mobilisation of our people's wisdom and resilience, based on the decades of tireless work done by AF-EC under its Director Dr Y.V. Malla Reddy's leadership, and which he has placed before us in his capacity as a co-author, we can overcome an existential threat to India's agriculture.

Former Secretary to the Prime Minister   K. R. Venugopal, IAS (Retd.)
Special Rapporteur to National Human Rights
Commission (NHRC)
Hyderabad, India
26 June 2020

Preface

This book is the outcome of a collaborative study undertaken by Livelihoods and Natural Resources Management Institute (LNRMI), Hyderabad, and *Accion Fraterna*-Ecology Centre (AF-EC), Ananthapuramu.

During the course of this study, we have received valuable support from various individuals and institutions, which ultimately paved the way for this book. We wish to acknowledge several of those who have provided invaluable support during the course of this study.

First and foremost, we humbly acknowledge the AF-EC, Ananthapuramu, for providing financial support to carry out this research work. Our sincere thanks are due to Shri K. R. Venu Gopal, IAS (Retd.), for kindly agreeing to write the foreword to this book. We would like to thank Dr. T. Yellamanda Reddy (Technical Director & Team Leader) and other Core Team/Executive Committee and Professional Team Members of AF-EC, particularly Mr. Bhaskar Babu, Mr. Naga Raju, Mr. Ramayyasetti, Dr. P. Lakshmi Reddy, Mrs. H. Rizwana, Mr. M. Shaikshavali, Mr. B. Fazlulla, and Field and Technical Staff of sample Watersheds, for all their support and encouragement during the period of this study. We would like to thank Mr. Y. Srinivasulu, field Supervisor, LNRMI, for efficiently organizing and coordinating field survey and data collection. Our thanks are due to Mr. Ch. Balaswamy for his support in data cleaning, tabulation and analysis. Dr. P. D. Sreedevi, National Geophysical Research Institute, Hyderabad, has helped in assessing the hydrogeology of the sample watersheds. Our thanks are due to her for the excellent work and the maps she provided.

Thanks are also due to all the research investigators for their dedicated work during field study. Last but not least, we would like to thank the farming communities of the sample villages who have been the main contributors to the study spending long hours in household survey, Focus Group Discussions and personnel interviews.

The communities' efforts and cooperation are gratefully acknowledged. Three anonymous reviewers of the publishers have provided constructive feedback on the draft outline and contents of the book. Their feedback, which helped in improving the outcome (book), is gratefully acknowledged.

Hyderabad, India
Anantapur, India
Hyderabad, India

V. Ratna Reddy
Y. V. Malla Reddy
M. Srinivasa Reddy

Contents

List of Boxes

List of Figures

List of Tables

Acronyms

1D	One Drought
2D	Two Droughts
3D	Three Droughts
ACIAR	Australian Centre for International Agricultural Research
AEI (Luxembourg)	*Aide à l'Enfance de l'Inde*
AEOs	Agriculture Extension Officers
AF-EC	*Accion Fraterna*-Ecology Centre
AI	Area irrigated
AP AGROs	Andhra Pradesh State Agro Industries Development Corporation Limited
AP	Andhra Pradesh
APDAI	Andhra Pradesh Drought Adoption Initiative
APDMP	Andhra Pradesh Drought Mitigation Programme
APE	agriculture productivity enhancement
APMC	agriculture produce marketing committees
APMIP	Andhra Pradesh Micro-Irrigation Project
APSAM	Andhra Pradesh State Agriculture Mission
arid	(<500 mm rainfall conditions)
ASMS	Apex Sasya Mitra Samakhya
ATL	Area Team Leader
BC	Backward Castes
BIFSRA	bio-intensive farming system in rain-fed agriculture
BLS	Baseline Survey
BoD	board of directors
BSL	baseline survey
BW	borewells
CB	capacity building
CBGWM	community-based groundwater management
CBOs	Community Based Organisations
CBP	Capacity Building Phase
CCTs	continuous contour trenches

CD	Check Dam
CIGs	common interest groups
C-IWMP	Control - Integrated Watershed Management Programme
C-NABARD	Control - National Bank for Agriculture and Rural Development
CSOs	Civil Society Organisations
CSR	corporate social responsibility
CW	Check Walls
CWB	crop water budgeting
CWS	Centre for World Solidarity
D.K.Thanda	Devankunta Thanda
DEM	Digital Elevation Model
DLH	Dryland Horticulture
DoA	Department of Agriculture
DoLR	Department of Land Resources
DoPs	Dugout Ponds
DoRD	Department of Rural Development
DPRs	Detailed Project Reports
DS	Downstream
EED (Germany)	Energy Efficiency Directive
EFMS	electronic fund management system
e-NAM	e-national agriculture market
EPA	Entry Point Activity
F	Female
FC	Financial Capital
FD	Fixed Deposit
FDI	foreign direct investment
FES	Foundation for Ecological Security
FFS	farmer field schools
FGDs	Focus Group Discussions
FIP	Full Implementation Phase
FPO	Farmer Producer Organisation
FPS	fair price shops
FPs	Farm Ponds
GCA	Gross Cropped Area
GIA	Gross Irrigated Area
GIS	Geographic Information Systems
GIZ	German Society for International Cooperation
GMG	Groundwater Management Group
GoAP	Government of Andhra Pradesh
GoI	Government of India
GO/Govt.	Government
GP	*Gram Panchayat*
GPMS	*Gram Panchayat* Management System
GS	*Gram Sabha*

GSA	Gross Sown Area
GSMS	*Gramasasya Mitra Samakhya*
GTZ	German Agency for Technical Cooperation
ha.	hectare
HC	Human Capital
Hect.	hectare
HEIDA	High External Input Destructive Agriculture
HH	household
HNSS	*Hundri Niva Sujala Sravanthi*
HRT	Horticulture
IC	Input Costs
ICCO	Interchurch Organization for Development Cooperation: Natural water courses
ICDS	Integrated Child Development Services
ID	irrigable dry
IFS	integrated farming system
IGAs	Income Generating Activities
IGWDP	Indo-German Watershed Development Programme
INM	Integrated Nutrient Management
IPM	integrated pest management
IWMP	Integrated Watershed Management Programme
KIIs	Key Informant Interviews
LE	livelihoods enhancement
LEISA	low external input sustainable agriculture
LF	Livelihood Fund
LH	livelihood
LL	landless
LMF	Medium and Large Farmers
LNRMI	Livelihoods and Natural Resource Management Institute
LV	Land Value
M-Book	Measurement Book
M	Male
MACS	Mutually Aided Cooperative Society
MDMP	Mid-Day Meal Programme
MDMS	Mid-Day Meal Schemes
Medi	Medium
Mega/Meso-Watershed	Watershed covering an area of up to 5000 ha. as against the earlier micro-watershed approach (500 ha.)
Micro-watershed	Watershed covering an area of 500 ha.
MF	Maintenance fund
MF	Marginal Farmers
MFTC	multiple fruit tree crops
M&E	monitoring and evaluation
MGNREGA	Mahatma Gandhi National Rural Employment Guarantee Act

MGNREGP	Mahatma Gandhi National Rural Employment Guarantee Programme
MGNREGS	Mahatma Gandhi National Rural Employment Guarantee Scheme
MI	Minor Irrigation
MMS	*Mandal Mahila Samakhya*
MoRD	Ministry of Rural Development
MoU	memorandum of understanding
MPIU	Mobile Protective Irrigation Unit
MPT	mini percolation tank
MPTs	mini percolation tanks
MRO	Mandal Revenue Official/Officer
MS	Midstream
MSMS	*Mandala Sasyamitra Samakhya*
MSP	Minimum Support Price
MWS	Mega Watershed
N. Puram	Narayanapuram
NABARD	National Bank for Agriculture and Rural Development
nala	Natural water courses
NC	Natural Capital
NCA	Net Cropped Area
NDDB	National Dairy Development Board
NFB	New Farm Bund
NGO	Non-Governmental Organisation
NIA	Net Irrigated Area
NI	Net Income
NPM	Non-Pesticidal Management
NREGA	National Rural Guarantee Act
NREGS	National Rural Employment Guarantee Scheme
NRM	Natural Resource Management
NSA	Net Sown Area
OC	Other Castes
OE	Over Exploitation
OPS	online payment system
OV	Output value
PC	Physical Capital
PDS	Public Distribution System
PFA	Project Facilitating Agency
PIAs	Project Implementing Agencies
PMKSY	*Pradhana Mantri Krishi Sinchayi Yojana*
PoP	poorest of the poor
PRA	Participatory Rural Appraisal
PRIs	*Panchayati Raj* Institutions
PSI	Productive System Improvement
PTs	Percolation Tanks

RARS	Regional Agricultural Research Station
RDT	Rural Development Trust
RF	revolving fund
RFCs	rain-fed farmers' cooperatives
RFDs	Rockfill Dams
RIDF	Rural Infrastructure Development Fund
RIDS	Rural Integrated Development Society
RO	Reverse Osmosis
RRA	Revitalizing Rainfed Agriculture
Rs.	Rupees
RSO	Resource Support Organization
RWH	Rain Water Harvesting
RWHS	Rain Water Harvesting Structures
S&M	Small & marginal
SA	Sustainable Agriculture
SB	Stone Bunding
SC	Scheduled Castes
SC	Social Capital semi-arid (>500 mm rainfall)
SES	Socio-Ecological System
SGWM	Sustainable groundwater management
SHGs	self-help groups
SLNA	State Level Nodal Agency
SMC	soil and moisture conservation
SMF	Small and Marginal Farmers
SMG	*Sasya Mitra* Group
SMGs	*Sasya Mitra* Groups
SNRM	sustainable natural resource management
SoI	Survey of India
SRGWM	Social Regulation In Groundwater Management
SRTM	Shuttle Radar Topographic Mission
SSWSS	*Sri Sathya Sai* Water Supply Scheme/Project
ST	Scheduled Tribes
STO	Socio Technical Organiser
TMC	thousand million cubic
TMCs	tank management committees
TRNC	Trenches
UGs	User Groups
US	Upstream
VDA	Village Developmental Activity
VDC	Village Development Committee
Velugu	poverty alleviation programme
VO	Village Organisation
VWDC	Village Watershed Development Committee
WASSAN	Watershed Support Services and Activities Network
WC	watershed committee

WDC	watershed development committee
WDF	Watershed Development Fund
WOTR	Watershed Organisation Trust
WS	Watershed
WSA	Watershed Association
WSC	Watershed Committee
WSCs	watershed committees
WSD	Watershed Development
WSDP	Watershed Development Programme
WSM	Watershed Management
WSMPs	watershed management programme
WSPs	watershed programme
WUAs	water user associations
Y	Yield

Chapter 1
Introduction

1.1 Background

South Asia is among the most affected regions due to climate variability, with increased frequency of droughts, increase in temperatures, shifts in rainfall pattern, intense rainfall, etc. (IPCC 2007).While agriculture as a whole is expected to be negatively affected, poorly endowed (low-rainfall and low-irrigation) regions in particular are expected to be impacted differently. Crop compositions and crop calendars might change with the pattern and structure of climatic variables. In the absence of accurate prediction and forecast of climate variability, farming communities and institutions are unable to prepare adequately to face the challenges and making them vulnerable. Vulnerability is a function of sensitivity, exposure, and adaptive capacity or resilience. Climate change or increased frequency of droughts influences the sensitivity. Climate change along with the geographical location influences the exposure and the ability of the household or communities in terms of natural, physical, financial, human, and social capitals determine their adaptive capacity or resilience. Besides, the degradation of natural resource base (land, water and biomass) is adversely affecting the resilience or adaptive capacity of these communities.

Given their geographical disadvantages, these farming communities are increasingly becoming sensitive, exposed, and less resilient to climate variability, i.e. more vulnerable over the years. While exposure to climate variability is geographical and long-term in nature, sensitivity is linked to exposure and intensity (severity) of the event. Resilience adoptive capacity, on the other hand, could reduce the impacts and can be dealt in the short to medium run. Resilience is defined as the magnitude of disturbance the socioecological system (SES) can tolerate and still persist (Holling 2001 as quoted in Carpenter et al. 2001). Another definition is the ability of the SES to resist disturbance and the rate at which it returns to equilibrium following disturbances (Pimm 1984 as quoted in Carpenter et al. 2001). While sensitivity and

© Springer Nature Switzerland AG 2020
V. R. Reddy et al., *Climate-Drought Resilience in Extreme Environments*,
https://doi.org/10.1007/978-3-030-45889-8_1

exposure adds to vulnerability, resilience reduces vulnerability through enhancing adaptive capacity of the household or community (Gallopín 2007).

Enhancing the biophysical attributes and natural resource base is critical for improving the resilience of agriculture or crop systems to climate variability. Watershed technology is seen as one of the best alternatives for improving the natural resource base through soil and water conservation techniques. It enhances the resource base benefiting agriculture in terms of reducing soil degradation, run-off, improved in situ soil moisture, access to irrigation, etc. and hence improves the resilience of the system. These impacts in turn could provide ecosystem services to downstream as well as urban and peri-urban areas in terms of drinking water, micro-climate, etc. At the same time, watershed interventions help improving the household wellbeing in terms of strengthening the financial, physical, human, and social capitals along with the natural capital. Thus watershed interventions could result in improving the resilience or adaptive of the resource base as well as the households or communities. On the other hand, watershed development (WSD) also could lead to over exploitation (OE) of groundwater, intensive agricultural practices, etc., which could adversely impact the natural capital. Some of the changes due to WSD work in the opposite direction in terms of improving the resilience of the SESs. That is, while WSD could enhance the resilience of one system (upstream/agriculture), it may adversely affect the resilience of another system (downstream/livestock).

Over the years, there has been a transformation in the approach to watershed programmes (WSPs) in India. WSD has transformed from being a mere soil and water conservation programme into a vehicle for transformation of rural livelihoods and development. While watershed interventions are viewed as necessary condition for enhancing the natural resource base and related benefits in the poorly endowed regions, they are not sufficient to sustain livelihoods and ameliorate farmer distress. This is increasingly becoming evident in the context of climate variability. Development practitioners working in these regions have realized that adaptation and mitigation strategies that can minimize climate variability or drought impacts (frequent and severe droughts) need to be identified. Organizations like Accion Fraterna-Ecology Centre (AF-EC) in Andhra Pradesh (AP) have realized that drought adaptation and mitigation strategies centring around watershed interventions would be effective in addressing the challenges of climate variability and drought. For instance, new strategies of land use, farming practices, water management practices, and income-generating and productivity-enhancing activities are introduced along with watershed interventions.

Over the years the planning, designing, and implementation of watershed management (WSM) have been undergoing changes to address the climate variability challenges. In order to minimize the adverse impacts, new technical and institutional interventions have been introduced. Besides, some of the state government initiatives and National Rural Employment Guarantee Scheme/Act (NREGS/NREGA) activities have been converged into watershed activities. These new approaches are expected to help improving the resilience of the system and reduce vulnerability of the communities. The effectiveness of these approaches needs to be assessed and scaled up as a drought adaptation strategy in the poorly endowed regions. The assessment would help in orienting watershed management programme (WSMPs) as a mitigation strategy by identifying appropriate interventions and prioritizing adaptation options. At the national/state level, the next transition of WSM should be towards making it as an effective resilience-building intervention strategy for climate variability/drought adaptation.

Though watershed is considered to be the best option for the poorly endowed regions for improving and stabilizing agriculture, watershed is yet to be recognized as a resilience-building option or adaptation strategy for climate variability/drought in India. Studies on resilience and vulnerability are mostly focused on technical aspects, by-passing the socio-economic dimensions (Ravindranath et al. 2011). Adaptation studies, on the other hand, deal with socio-economic aspects at the household level trying to understand the strategies households adapt to adjust to the impact of climate variability without considering the technical information on temperature, precipitation, hydrogeology, etc. (see for review Reddy et al. 2010). Of late integrated research has been initiated where multidisciplinary approaches like agrometeorology, hydrology, soil sciences, and socio-economic aspects are being adopted (see for details, Reddy and Syme 2015; ACIAR 2015). However, these studies are constrained by lack of real time on ground evidence, as they draw from the technical (scientific) aspects of the interventions and their likely impacts on natural resources. Besides, these studies are based on the experience of watersheds implemented earlier, i.e. before Integrated Watershed Management Programme (GoI 2008). Besides, these watershed interventions have no specific focus on drought resilience or climate/drought proofing.

This book assesses the WSD interventions along with the complementary adaptation strategies introduced by the AF-EC in Ananthapuramu district of AP, which represents the poorly endowed harsh environments in India. AF-EC has been implementing watershed programme over the last three decades and highly regarded for its efficient and effective implementation at the state and national levels. While AF-EC recognizes WSD as a necessary condition for improving agriculture in these naturally disadvantaged regions, it continues to focus on further strengthening the watershed interventions in this region in order to make farming systems drought- and climate-resilient. In the process AF-EC team emphasizes on evolving new interventions and support mechanisms in the form of more appropriate and technically sound interventions, new cropping and farming systems (including livestock and allied activities), livelihood-focused institutional arrangements, etc. Effectiveness of these interventions is assessed in the context of two different watersheds supported by the state government (IWMP) and National Bank for Agriculture and Rural Development (NABARD). These interventions need to be highlighted for drawing lessons for watershed implementation elsewhere in the state and the country. Besides, a proper assessment could help in identifying the gaps and taking up corrective measures and improvisations.

1.2 Ecological Conditions of the Region

Watershed interventions need to be viewed in the light of harsh environmental conditions of the region. For, the biophysical conditions, rainfall and soils are main limiting factors for effective impacts of the watershed interventions. The region is located at the heart of *Deccan* Plateau (in South India) is arid, semi-arid, and chronically drought-prone for centuries. This backward region has a dubious distinction for twin problems of drought and poverty. It is the second driest region in India after Rajasthan. This geopolitical region consists of four districts, i.e. Ananthapuramu, Cuddapah, Chittoor, and Kurnool. The region has a geographical area of 6.7 million ha and has a population of 13 million. The cultivated area is 2.4 million ha of which only 0.56 million ha (23%) area is irrigated with groundwater and some surface water. It has 1.5 million ha (22%) under forests mostly without tree cover. The sources of irrigation (groundwater as well as surface) are undependable in the region. There is very little industry in the region, and livelihoods are dependent on mostly rain-fed farming which is prone to frequent droughts. The annual average rainfall in this region ranges from about 350 mm to about 650 mm from both southwest monsoon (from June to September) and north-east monsoon (from October and November). From the year 1876 till 1975, in 100 years 64 years received less than normal rainfall and had witnessed more than 50 drought years including severe famines. In 1876 the region experienced a severe famine wherein it was believed that almost 40% of the population died of hunger.

The last 20 years, from 1995 the region has experienced an increased frequency of droughts. In February 2012, the district administration of Ananthapuramu district reported that there were only 2 normal crop years in the past 14 years.[1] There is a clear shift in the onset of sowing season, and the dispersion of rainfall has further increased. The crop-season (July–October) rainfall decreased, and outside crop-season rain has been increasing, whereas, the volume of rainfall has not increased. The number of rainy days (rain events with a volume of >10 mm) has decreased from about 35 in 1980s to about 25 in the decade of 2010. At the same time, the intensity of rainfall in an event has increased. The duration of the dry spells in rainy season and the scantiness have increased. Temperature variability is increasing. Farmers say that the gap between day temperatures and night temperatures is widening. The summer is setting in as early as February. The temperature fluctuations during the day increase like a sudden rise in the temperatures in the evenings.

The above changes in the rainfall and temperatures have severe impact on crop production, livestock, and livelihoods of the fragile arid region. The major rain-fed crops in the region include groundnut, jowar, bajra, red gram, cowpea, and Bengal gram. The duration of these crops ranges between 100 and 160 days. They require timely distribution of rainfall of about 20 mm each at intervals of about 10 days during the crop period. One dry spell of over 30 days could destroy the crop. In the past 20 years, such long dry spells of more than 20–30 days are on the rise causing

---

[1] For detailed account of the climatic conditions in Ananthapuramu, see Reddy 2017.

crop failures. Distribution rather than volume of rainfall is very important for these crops. Further the temperature variability is said to be adversely affecting the crop production. Farmers say that earlier droughts were associated with failure of rains. But now droughts occur because of high and fluctuating temperatures. This is reflected in the dismal record of 2 good crop years in 14 years since 1998.

The absence of timely rains coupled with increased temperatures and pressure on groundwater has increased resulting in over exploitation. This has led to depletion of groundwater table and drying up of bore wells. Even the high-value horticulture crops like sweet lime, citrus, pomegranate, banana, and papaya are wilting due to drying up of bore wells. Due to failure of these high capital intensive crops coupled with the sunk costs in bore wells, farmers get into indebtedness. The increased dispersion of rainfall is also affecting the surface water storage systems, i.e. traditional surface water bodies. Earlier, these storage systems used to fill up at least once in 3 years. They are now filling up only once in 10 years resulting in poor groundwater recharge. The traditional water bodies used to provide drinking water for livestock earlier. Now most of the time they are dried up, and the cattle have to tread long distances to quench their thirst.

Earlier many farmers used to have livestock as part of their farming system. It was farmers' main coping mechanism from droughts. The crops they used to grow were mostly dual purpose for grain and fodder. With crops failing more often, the fodder is scarce now. The traditional pastures are now barren. Distress sale of livestock is widespread, especially during the severe drought years. Due to the increased intensity of rainfall, the soil erosion has increased, and the biomass in the soil (soil organic carbon) has been coming down faster. The land is more exposed to sun, wind, and rain affecting the flora and fauna and as well the biotic life in the soil. The degradation process was aggravated by decades of monocropping, particularly of groundnut, which does not leave any crop residue in the soil. Increased soil erosion has led to siltation of local surface water bodies reducing their storage capacity.

The above changes have affected seriously the farming system and thrown the farmers into severe distress. The impact has been even more telling on the fragile, small, and marginal farmers. With virtually no other nonfarm livelihoods, Ananthapuramu backwardness and poverty are well reflected in its severe rural indebtedness, high participation rates under Mahatma Gandhi National Rural Employment Guarantee Scheme (MGNREGS), and rampant farmers' migration including seasonal. The cascading effects of all these pushed farmers into distress and forcing farmers to commit suicides due to social pressures. There were about 3000 farmers' suicides reported during the past 20 years, as 90% of the farmers have been trapped in deep debts and see no hope in farming. The agricultural crises and rural distress deepened after the advent of economic liberalization during the 1990s. Reforms led to increased crop investments due to cuts in fertilizer subsidies and increase in other input costs with no matching increase in output prices. At the same time the living costs, health, and education costs have gone up substantial. The high input cost agriculture is biased in favour of rich farmers with irrigation facilities making small and marginal rain-fed farms un-viable (Reddy 2017; Reddy and Reddy 2017).

Though watershed-centred interventions help in enhancing the conditions in a normal situation, they fail to sustain the effectiveness of the interventions in harsh conditions as narrated above, even in the medium term. These harsh environmental conditions are fostered with climate change-/drought-related risks. For one drought can wipe out all the gains. Households could hardly withstand one drought and often abandon farming in the event of consecutive droughts. Though land use or cropping pattern changes along with technological options could address these problems to some extent their effectiveness is limited in the context of increasing water stress. In this context there is need for finding long-run solutions like enhancing the water resource potential of the region, conjunctive use of water sources (surface and groundwater), groundwater management, and equitable distribution of water. Though these aspects are in the purview of policy and administrative interventions, they are mandatory for ensuring sustainable livelihoods and address farm distress in a systematic manner. Some of these options are identified in the process of our assessment of the interventions and their impacts.

1.3 Watershed Development Programmes (WSDPs)

A number of agencies implement watershed development programmes in India. Of these the programmes implemented by the Government of India (GoI) and the National Bank for Agriculture and Rural Development (NABARD) are prominent. Integrated Watershed Management Programme (IWMP) of the GoI and the NABARD implemented watershed development programme and adopted different approaches in terms of scale, implementation process, and livelihood components. The salient features of these two programmes are discussed below.

1.3.1 Integrated Watershed Management Programme (IWMP)[2]

Integrated Watershed Management integrates natural resource management (NRM) with community livelihoods in a sustainable way. It addresses degradation of natural resources, viz. soil erosion, floods, frequent droughts and desertification, poor water quantity and quality along with low agricultural productivity, and poor access to land and related resources from an integrated watershed management perspective. It also addresses the new and upcoming challenges and opportunities for the region and climate change at the watershed level. Gram Panchayat (GP) has been effectively involved at village/watershed level to keep transparency and people's

---

[2] Drawn from GoI (2019a, b); Department of Land Resources, Ministry of Rural Development, Govt. of India (https://dolr.gov.in/). GoAP (n.d.), Watershed Development (PMKSY–Watersheds): Department of Rural Development (http://iwmp.ap.gov.in/WebReports/Content/Programmes. html) (https://core.ap.gov.in//CMDashBoard/UserInterface/IWMP/IWMPREPORT.aspx).

participation. GP supervises, supports, and advises watershed committee (WC) and authenticates accounts/expenditure of WC and other institutions of watershed projects. It facilitates convergence of other programmes, maintains asset register to retain it after the project, provides office accommodation and other requirements to WC, and allocates usufruct rights to deserving user groups (UGs)/self-help groups (SHGs) over the assets created.

IWMP was introduced during 2008–2009 with new guidelines from the central government. IWMP has introduced the concept of *mega-/meso-*watershed covering an area of up to 5000 ha as against the earlier micro-watershed approach (500 ha). The increased scale is expected to enhance the effectiveness of the economic impacts due to more comprehensive environmental benefits associated with the scale. On the other hand, the increased scale of operation is expected to complicate the collective institutional aspects such as promoting watershed associations and committee's due to the involvement of multiple communities across villages. To overcome this, most states have adopted the cluster approach of creating micro-watershed institutions at the village level and providing village wise design and implementation (DPRs), while keeping the mega watershed in view. IWMP guidelines also gave more emphasis on strengthening the livelihoods with allocation of more funds (9%). Besides, productivity enhancement interventions get another 10% of the allocations. Over all, allocation per unit of land has also been enhanced to 12,000 per ha. The first batch IWMP watershed was initiated during 2009–2010. Muttala watershed from the first batch has been selected for this study.

1.3.2 NABARD Watersheds[3]

NABARD started implementing participatory watershed projects under Indo-German Watershed Development Programme since the 1990s. Based on the success of the earlier programmes, NABARD has constituted a Watershed Development Fund (WDF) to spread and promote participatory watershed development models and to create the necessary framework conditions to replicate and consolidate isolated successful initiatives under different programmes in the government, semi-government, and the non-governmental organization (NGO) sectors. With the advent of corporate social responsibility (CSR) in India, corporate bodies are also engaged in implementation of watershed projects on a co-funding basis.

Watershed Development Fund (WDF) was created in NABARD in 1999–2000 with an initial corpus of Rs. 200 crore. The corpus was augmented over the years by the interest differential earned under rural infrastructure development fund (RIDF) and the interest earned accrued on the unutilized portion of the fund. The financial assistance under the programme was in the form of grant or grant-cum-loan. As on

---

[3] Drawn from: (https://www.krishaksarathi.com) (https://www.krishaksarathi.com/watershed-development-programme.html; https://www.nabard.org/demo/auth/writereaddata/File/21%20_WATERSHED_MANAGEMENT.pdf).

31 December 2017, the total corpus stood at Rs. 1174.82 crore with cumulative number of projects of 600. The projects covered an area of 5.94 lakh ha in 18 states, with a commitment of Rs. 417 crore. Main objectives of the programme include soil and water conservation, climate change adaptation and mitigation, effective and sustainable use of available water resource, enhancement of farm production, productivity and income of farmers, improvement of skill and employment opportunities in the area, creation of additional employment potential for the small/marginal farmers and agricultural labourers, and improvement the socio-economic status of the farmers.

To address the sustainable development in completed watershed projects, NABARD has come out with sustainable development policy, wherein, the critical issues of technology transfer, agriculture extension, credit intensification, integrated pest management (IPM), integrated nutrient management (INM), promotion of Farmer Producer Organisations (FPOs), etc. are being initiated in the post-watershed development period through capacity building and leadership development of the watershed community.

NABARD adopts the Indo-German Watershed Development Programme (IGWDP) approach supported by its WDF across the country with its own guidelines. It adopts a two-phase approach, viz. capacity building phase (CBP) and full implementation phase (FIP), which is very close to AF-EC approach. NABARD guidelines emphasize components like livelihood support activities especially for the landless by supporting off-farm and nonfarm livelihood activities. Formation of Mutually Aided Cooperative Societies (MACS) for post-project sustainable management of livelihood funds as well as watershed assets, which is unique. Besides, it also has a component for demonstration on 100 ha to showcase watershed interventions and encourage participation. It also promotes and integrates specific interventions like drought and climate-resilient crop systems. NABARD watersheds are adopted at about 1000 ha scale. A NABARD watershed implemented in one of the lowest rainfall (about 400 mm) Mandals (Kalyandurg) in the country has been selected for assessment.

1.3.3   AF-EC: Implementing Agency[4]

AF Ecology Centre was founded by Father Vincent Ferrer in 1982. Since then it has been involved in rural development particularly in drought mitigation, watershed development, environmental development, and policy advocacy. AF-EC has made a substantial contribution to Ananthapuramu district since 1986 with its participatory watershed development programme supported by EED (EED) (Germany) and ICCO (ICCO) (Netherlands). It was perhaps the largest participatory watershed programme by an NGO in India spread over about 300 villages, covering about 1.35

---

[4] Office of the Director, Accion Fraterna – Ecology Centre, Ananthapuramu.

lakh ha of farm land and 60,000 farmers. It was known for its participatory approach and very high-quality watershed implementation on a sizable scale. The major interventions under the watershed programme include soil and moisture conservation, rainwater harvesting, horticulture, rain-fed agronomical practices, and biogas and people's institutional development.

Integrated watershed development interventions and promoting village-level community institutions started emerging as focus for AF since 1996. "Village watershed" concept through which the entire land in the village as micro-watershed was taken up for integrated watershed development in a participatory approach. This micro-level and integrated ridge to valley participatory watershed development approach in selected villages has created better impact in terms of enhancing natural resource endowment and increasing carrying capacity of natural resources in watershed villages and also organizing village-level community institutions at village level for sustainable management. In 2002 it was separated from Rural Development Trust (RDT) and made an autonomous organization with an independent office and campus. The idea was to decentralize and specialize in ecology and environment. However, it remains and continues to be a sister organization of RDT and retains its roots and motivation.

The next phase in the growth of the new autonomous organization went a step further to focus more on developing "model watershed villages," a concept that went beyond watershed activities in order to mobilize and organize the human and institutional resources in a village for its holistic and sustainable development. It was in this phase that AF-EC actively engaged in strengthening village-level watershed institutions like *Gram Sabha* (GS), watershed development committee (WDC), and village development committee (VDC) and linking them with other village-level institutions like GP, water users association, livestock centres, schools, primary health centres, etc.; AF Ecology Centre was working in about 100 villages with a Model Village Watershed Development approach spread over 15 Mandals of Ananthapuramu district.

AF-EC shifted its programme focus from watershed based to sustainable agriculture in the year 2007. This approach integrates the following key aspects:

- Promoting vibrant people's institutions like *Sasya Mitra* Groups (SMGs) in order to actualize their own potential and access opportunities with government banks, private sector, NGOs, etc., gender and social equity are an integral part of people's institutions.
- Developing models of integrated farming systems and strategies for coping with droughts and enabling livelihood security for rain-fed farmers. It includes rain-fed agriculture, on-farm, off-farm, and nonfarm interventions.
- Demonstrating effective proven and scalable practices, technologies and models at a visible scale to the farmers, government functionaries, policy makers, etc.
- Public opinion building and policy advocacy for pro-poor and pro-environment policies and programmes.

However, it continued to implement participatory watershed development projects with the support of NABARD and IWMP (Integrated Watershed Management

Project) and MGNREGS of GoI. It has also made a significant contribution in influencing a favourable and enabling policy conditions for a people-centred watershed development and rural employment (MGNREGS) in the state of Andhra Pradesh. At the policy level, AF-EC has been actively involved in various policy making bodies like Andhra Pradesh Water Conservation Mission, Andhra Pradesh State Commission on Farmers Welfare, and Advisory Committee on Watershed Development Programme of Andhra Pradesh. Presently AF-EC is a member in Andhra Pradesh State Agriculture Mission (APSAM) a body to advise the Govt. of AP on policies of agriculture and farmers' welfare. Further AF has been actively involved in various consultations by the Ministry of Rural Development (MoRD) at national level.

AF-EC adopts the same strategy for all the watersheds implemented by them, which include IWMP, NABARD, and their own watersheds. Except for some minor deviations across funders, the philosophy is same across the watersheds.

1.4   Objectives and Approach

The book assesses and evaluates two major watershed interventions implemented by the AF-ECs in Ananthapuramu District. Specific objectives include the following:

1. Assess the impact of watershed interventions on crop production.
2. Evaluate the livelihood impacts on the communities.
3. Assess the drought and climate resilience of the households in the context of watershed and related livelihood interventions.
4. Draw lessons and identify measures to strengthen and improvise the interventions for enhanced drought and climate resilience.

The assessment would provide insights into the way AF-EC implemented IWMP/NABARD watershed programmes and the impact of these watersheds, especially in the harsh climatic conditions. It helps a comparative assessment of IWMP and NABARD watersheds in terms of ecological (scale), socio-economic, and institutional aspects. For this is one of the first systematic evaluations of IWMP in the country along with the NABARD watersheds, especially under best implementing conditions. The assessment is expected to further strengthen the AF-EC interventions in enhancing the drought resilience of the households and farming systems.

1.4.1   Evaluation Framework

Sustainable development has brought into focus the nexus between human (socio-economic) and natural (environmental) systems. Integrating environmental concerns into evaluation studies has become critical for moving towards achieving sustainable development goals (Uitto 2019). Accounting for environmental sustainability could drastically alter the evaluation outcomes and policy process. Incorporating the environmental impacts in project evaluation helps in internalizing the externalities associated with resource use, development, and degradation. Benefit flows depend on the nature of externalities (positive or negative). At the same time, externalities could be pervasive spatially as well as temporally. This is more so in the context of NRM, viz. land and water resources. For instance, land use changes (cropping/fallowing) could influence runoff and recharge of precipitation in the short run or increase the precipitation in the long run (increase in forest cover).

Improved recharge or reduced runoff in the upstream areas due to watershed interventions could improve groundwater table in the downstream areas. Besides, interventions in the upstream could provide number of ecosystem services like quality of drinking water. On the contrary, watershed interventions like water-harvesting structures in the upstream could reduce water flows into the downstream water bodies and adversely impact groundwater recharge as well. Similarly, unsustainable farm practices (excess use of chemicals) in the upstream could adversely impact the ecosystem services like water quality in the downstream. Evaluation of watersheds considering only the on-site benefits fails to account for such benefits/costs and hence underestimates/overestimates the benefit flows. Therefore, investments in upstream areas cannot be justified by their on-site benefits alone and can only pass economic reasoning when downstream benefits are embodied.

Research has revealed that the micro-watershed approach may be producing hydrological problems that would be best addressed by operating at a macro-watershed scale. For example, in India, recent hydrological research cautions that watershed projects may be aggravating precisely the very water scarcity they intend to overcome. Batchelor et al. (2003) noticed that successful water harvesting in upper watersheds came at the expense of lower watershed areas. This indicates that watershed interventions check the movement of both surface runoff and groundwater towards downstream locations. Agriculture intensification associated with watershed interventions are leading to water quality problems (eutrophication) in the medium to long run, especially in the groundwater-dependent regions (Reddy et al. 2018a, b). Perpetuation of such problems could jeopardize food security in the long run. This indicates two adverse project outcomes: first, what is good for one micro-watershed can be bad for others in the downstream locations and, second, what is good for a watershed in the short term can be bad in the long term. Thus, evaluation studies should consider spatio-temporal externalities associated with natural resource use in order to move towards achieving sustainable development.

The increased scale of watersheds brings in the advantages as well as disadvantages as far as the effectiveness of the program is concerned. IWMP watersheds at the 5000 ha scale should help internalize the externalities associated with hydrogeology and biophysical aspects. On the other hand, it could hinder the institutional aspects pertaining to collective strategies. Hence, it is necessary to assess the impacts of watershed interventions using an integrated approach using an appropriate evaluation framework.

In this book, the integration is mainly in terms of biophysical and socio-economic and environmental aspects (Reddy and Syme 2015) of the watersheds at a scale of

1000 ha and above. The biophysical aspects include hydrogeology, rainfall, and land use, while the socio-economic and environmental aspects incorporate household resilience in relation to its livelihood capitals (environmental and socio-economic). The integrated approach provides an insight into the interactions between the hydrogeological and biophysical aspects of a watershed and the resulting influence on the quality and quantity of watershed impacts on the livelihoods of the local communities. Besides, it also explores the potential of watershed development in the context of increasing climate variability as a mitigation or adaptation strategy for improved resilience of the farming communities.

The approach highlights the importance of understanding these complex interactions specifically in the context of harsh environments and their importance in achieving not only sustainable soil and water management but also economic and livelihoods outcomes. Further, it also assesses the role of interventions that complement the watershed interventions to address the drought/climate impacts in the region. These include land use changes, water management strategies, and water and moisture conservation technologies. In the process equity impacts of all the interventions are assessed, and the need for stakeholder engagement at all levels if the integrative approach is to be useful for meaningful evaluation of watershed programs at different levels.

The integrated approach is primarily driven by the socio-economic aspects. Watershed impacts are assessed in terms of household resilience to climate/drought impacts. The level or degree of resilience (number of droughts a household can withstand) varies across households. The degree of household resilience is linked to the household's assets and capabilities. Sustainable livelihoods (five capitals) framework provides deeper understanding of household assets and capabilities and their role in mitigating vulnerabilities or enhancing the resilience (Reddy and Syme 2015).

The biophysical aspects influence household assets and capabilities through its natural capital, especially water and land. Biophysical attributes, including hydrogeology, rainfall, and soil type, are exogenous or given to the household and need to be taken into account while assessing the watershed impacts. These attributes are critical in determining the extent of impacts and should be considered while designing interventions in order to optimize the impacts. Of these, hydrogeology and land use are highly variable and instrumental in creating inequity in access to resources, assets, and capability. However, some households could substitute the lacuna in these attributes with other capabilities (capitals) such as human or social capital in order to enhance their resilience.

The integrated approach is based on the research that has adopted a clear analytical framework and scientific approach for assessing the watershed impacts. This approach provides a comprehensive assessment of sustainable watershed interventions and their impacts on livelihood outcomes of the communities. The hydrogeology assessment of the watersheds includes groundwater, surface-subsurface water, and land use assessments. These assessments are used to arrive at appropriate watershed intervention, viz. water-harvesting structures like check dams that are location-specific. The nature and density of such interventions are determined by exogenous factors including rainfall, soil quality, slope, aquifer structure, and land use (forests, wastelands, etc.).

The cropping pattern in a specific area influences the groundwater use and balance; and crop patterns are sustainable when crops are grown according to these biophysical attributes – that is, when crops are chosen according to the soil type and groundwater potential (sustainable groundwater yields), it can be termed as a sustainable crop pattern. Community livelihoods are determined by the biophysical potential of the region that can support farm systems. While agricultural or farm systems could enhance livelihoods in terms of financial capital, there are other forms of household assets and capabilities (human, physical, and social) that could potentially enhance livelihoods. Watershed interventions might directly or indirectly influence these capitals through strengthening of natural capital.

Hence, watershed impact assessments should look beyond natural and financial capital on which watershed interventions have a direct bearing. The socio-economic approach adopted here looks at the five capitals and the capabilities of the household, along with a number of indicators of these five capitals including the biophysical aspects to explain the variations in watershed impacts (resilience) between upstream/downstream and control situations.

Equity is assessed in terms of horizontal and vertical distribution of benefits. Horizontal equity is assessed by comparing upstream/downstream impacts, while vertical equity is assessed in terms of distribution of benefits within upstream/downstream locations. The integrated model helps in assessing whether the distribution of benefits is optimum, given the biophysical attributes of the specific location. It helps in arriving at alternative and appropriate design interventions that could optimize the benefits. Equity would be optimum when benefits are maximized across locations and sections of population. Maximal equity is not necessarily the absolute equity, which is ideal and desirable, and appropriate policies (compensation, subsidies, incentives, payments for environmental services, etc.) could help improve equity to a large extent.

Groundwater is the dominant source of irrigation in the rain-fed regions and plays a key role in the socio-economic development in agrarian economies. Since groundwater remains hidden in a complex system of rocks, its precise assessment is difficult, and this has resulted in a large mismatch between groundwater demand and availability. While watershed interventions are expected to improve groundwater recharge through better soil and water conservation practices, the actual availability of groundwater for final use depends on the suitability of interventions to the aquifer system.

In the context of watershed interventions, it is often presumed that groundwater recharge improves as one moves from upstream to downstream locations. However, these observations are not based on scientific information on aquifers and drainage systems, and often efforts on WSD go to waste. Thus, hydrogeological investigation through geophysical methods provides a clear link to the socio-economics since a precise knowledge of the subsurface is helpful in two ways: to ensure efficacy and suitability of the type of the WSD and to plan for its sustainable use. Further, differences in the nature and type of aquifers across the locations within a hydrological unit could result in contradictory evidence; and in the absence of such scientific information, watershed impacts have been attributed to the quality of implementation. Hence, there is a need to understand the role of hydrogeology in terms of water resource potential and its use in the context of watershed interventions.

1.4.2   Approach and Methodology

The book has assessed and evaluated one each of IWMP and NABARD watersheds implemented by AF-EC. While NABARD watershed covered one village, IWMP consisted of four micro-watersheds covering four villages. The sample watersheds were selected from three IWMP watersheds and six NABARD funded watersheds implemented by AF-EC. IWMP watersheds cover 2500–5000 ha. NABARD watersheds cover about 1000 ha. Besides two control villages were selected for each of IWMP (C-IWMP) and NABARD (C-NABARD) watersheds. Given the objectives, the assessment has been carried out from multiple angles, viz. socio-economic, environmental, technical, and institutional. Socio-economic aspects were assessed in terms of economic benefits in terms of crop yields, returns to agriculture, other income-generating activities (IGAs), education, social and gender equity, etc. Environmental benefits are assessed in terms of improved natural resource conditions, viz. hydrogeology, land and water, resilience, etc. Technical aspects like new technologies, techniques, crop systems, cultivation methods, etc. were assessed. Institutional aspects such as watershed committees (WSCs), UGs, SHGs, etc. were assessed. All these aspects were evaluated for impacts (positive/negative) and their sustainability over the years.

The sustainable livelihoods framework (five capitals) is adopted to get a better understanding and insights into the livelihood aspects of the households in the context of watershed and other related interventions. The five capitals (natural, physical, financial, human, and social) are considered to assess the household's strengths, abilities, and constraints in dealing with drought and climate variability challenges. Household perceptions about the present status of these five capitals and their contribution in enhancing their resilience are elicited. And which capital they need to improve in order to make them more resilient is also assessed. Each capital is represented by various indicators, viz. natural capital, land, water, etc.; physical capital, livestock, implements, machinery, etc.; financial capital, income, savings, debt, etc.; human capital, education, skills, health, etc.; and social capital, groups, administrative and political connections, etc.

Both qualitative and quantitative research methods were used for this purpose. Qualitative research tools such as focus group discussions (FGDs), key informant interviews (KIIs) and discussions, and case studies were used to elicit information at the community, village, and cluster level. The externality impacts of these programmes can be captured better at the broader community level rather than at the individual household level. KIIs with the members of user groups and village elders were conducted. Check list of questions were prepared for eliciting the information

Table 1.1 Details of qualitative research

Village	FGDs	Group type	KIIs	Personal
Muttala (IWMP)	4	Mixed groups: WSC (four micro-watersheds), watershed sub-committees/UGs, GP, SHGs, VO, irrigated and unirrigated farmers and labourers	5	Director, Assistant Director, ATL, Project Officer, and STO
Battuvanipalli (NABARD)	4	Mixed groups: VDC, GP, WSC, MACS, SHGs, VO (executive committee and *Velugu* animators), farmers (irrigated and unirrigated) and labourers	5	Director, Assistant Director, ATL, Engineer and STO
Kurlapalli (C-IWMP)	3	Village institutions, farmers, and labourers	–	–
N. Puram (C-NABARD)	3	Village institutions, farmers, and labourers	–	–

Note: FGDs Focus group discussions, *KIIs* key informant interviews, *C-IWMP* control village of IWMP watershed, *C-NABARD* control village of NABARD watershed

from the stakeholders. The details of qualitative research conducted across the watershed and non-watershed villages are presented in Table 1.1.

Quantitative information was collected from the secondary as well as primary sources. Secondary information pertaining to the programme design, coverage, and achievements (physical and financial) were collected from AF-EC documents at the watershed and village levels. Information was also drawn from the baseline survey (BLS) conducted by the implementing agency (AF-EC) at the beginning of the project. Household-level (including farm level) primary information was collected with the help of a structured questionnaire using appropriate and scientific sampling methods. Socio-economic information along with the livelihoods information at the household level was collected. The sampling details are presented in Table 1.2. The micro-watersheds of the IWMP watershed represent upstream (ridge) and downstream (valley) locations of the watershed. Samples of 50 households were drawn from each micro-watershed. Minimum sample of 50 is adhered to in order to ensure statistical robustness of the assessment. Sample is proportionately drawn representing all the four major economic classes, viz. landless (LL), small and marginal (SMF), medium- and large (LMF)-category farmers. Further, care was taken to include households from different social groups, i.e. scheduled castes/tribes, backward, and others. Altogether, samples of 350 households were drawn, accounting for 20% of the actual population, though it varies across the villages. However, the sample distribution is more consistent across economic categories (Table 1.2).

Various aspects of watershed and related interventions are assessed using before and after as well as with and without watershed contexts. Before and after information was elicited from the households through requesting the respondents to recall the pre-watershed situation. This approach is usually constrained by the memory lapse of the respondents. But, memory lapse may not be a serious issue here due to the shorter time lag, i.e. less than 10 years. Besides, with/without approach of

Table 1.2 Sampling details of selected watersheds

Watershed	LL Actual	LL Sample	S&M Actual	S&M Sample	Medium Actual	Medium Sample	Large Actual	Large Sample	All Actual	All Sample
D.K. Thanda (US)	15	10	60	18	58	17	15	5	148	50 (34)
Goridindla (MS)	39	10	113	20	89	16	25	4	266	50 (19)
Muttala (MS)	58	10	149	16	141	15	81	9	429	50 (12)
Papampalli (DS)	60	10	41	15	41	15	30	10	172	50 (29)
Battuvanipalli	54	10	77	19	27	7	56	14	214	50 (23)
Kurlapalli (C-IWMP)	10	10	44	22	16	13	8	5	78	50 (64)
N. Puram (C-NABARD)	85	10	235	26	77	8	31	6	428	50 (12)
Total	**321 (19)**	**70 (20)**	**719 (41)**	**136 (39)**	**449 (26)**	**91 (26)**	**246 (14)**	**53 (15)**	**1735 (100)**	**350 (20)**

Note: Figures in brackets are respective percentage to the actual population
US Upstream, *MS* midstream, *DS* downstream, *LL* landless, *S&M* small and marginal
C-IWMP control village of IWMP watershed, *C-NABARD* control village of NABARD watershed

comparing the watershed villages with the non-watershed (control) villages would help reducing the memory lapse problems, while the later approach is constrained by non-availability of perfect matching village for comparison. Netting out the impacts by adopting both the methods together could address the problems to a large extent. The method of using both the methods (before/after and with/without) is known as "double difference" method. The means "t" test is used to test the robustness (statistical significance) of the impacts.

Watershed is a hydrological unit, and the effectiveness of its interventions largely depends on the geohydrology of the location. This is more so in the groundwater-dependent regions like Ananthapuramu. For improvement in groundwater potential depends on aquifer geometry, i.e. shallow, deep, fractured, withered, etc. In fact, watershed interventions need to be designed as per the aquifer characters in order to optimize the recharge benefits. In order to assess the suitability of watershed interventions, hydrological mapping has been carried out for the selected watersheds. The mapping helps in assessing the groundwater potential and type of interventions required at various locations. The micro-watersheds within the IWMP cluster are hydrologically determined as upstream (US); midstream (MS), and downstream (DS). This helps in assessing the effectiveness of interventions across the streams.

1.4.3 About the Book

This book is based on the collaborative research of Livelihoods and Natural Resource Management Institute (LNRMI) and AF-EC on assessing the potential of the watershed and watershed-centred interventions in addressing climate and drought risks. The book has adopted an integrated approach in assessing the impacts of watershed and its complementary interventions in two different watersheds situated in a low-rainfall and resource-poor region. It explores the generality of the approach taken by AF-EC in addressing climate/drought risks in harsh environmental conditions. The nexus framework of environmental, socio-economic, and institutional aspects is adopted to move towards sustainable resource management and livelihoods. Groundwater management is a critical issue in these climatic conditions. How sustainable management of the natural resources not only ensures improved resilience against drought, but also equity across locations and socio-economic groups is explored. Besides, the relevance of the sustainable livelihoods approach in the context of resilience in differing socio-economic and demographic conditions and alternative water and land management institutions are examined. The book is organised in six chapters (including the introduction chapter).

The objectives of the book along with the evaluation framework and approach are discussed in detail in the introductory chapter (one). The analytical framework discusses the importance of the integrated approach adopted that helps in internalizing the environmental externalities associated with natural resource use and exploitation. It argues that such an integrated approach to evaluation helps in achieving the sustainable development goals. An overview of the approach and sampling methodology is provided in the introductory chapter (one).

Biophysical and hydrogeology are region- and location-specific. These characteristics determine the exposure to climate/drought risks of the communities and households. Besides, they also play a critical role in influencing the effectiveness of watershed interventions. It is often assumed that watershed interventions improve the availability of water resources. But the magnitude of such improvements varies across agroclimatic contexts with differential rainfall and aquifer characteristics. Performances of watershed interventions are observed to be subdued in low-rainfall (arid/semi-arid) regions. The impacts of watershed interventions in enhancing groundwater recharge are limited in shallow aquifers. Similarly, socio-economic characteristics of the communities determine their vulnerability and adoptive capacity to climate/drought risks. Detailed profiling of socioecological characteristics along with hydrological mapping is critical for evaluating the performance of the interventions and institutional evolution in the selected watersheds and villages (Chap. 2).

Proper implementation of the watershed development programmes and the quality of interventions are critical for sustain the impacts in the long run. This is more so in the context of climate/drought risks. For communities see the benefits of interventions only when rainfall is good, i.e. once in 5 years in the study regions. Besides, of late rainfall intensity is raising that too during non-season. Under such conditions taking care of the structures requires lot of commitment and capacities among the communities. The implementation process along with the watershed-related interventions, including technologies and techniques, institutional evolution in the watershed villages, and awareness and capacity building initiatives, is discussed in Chap. 3.

Watershed and related interventions are aimed at improving the livelihoods of the communities and enhancing the resilience of the households against drought/

climate risks. The five capitals' approach of sustainable livelihoods framework has been used to assess the impact of the interventions. The resilience of the households is measured in terms of number of droughts a household can withstand. Resilience of the households is linked to the status of five capitals of the households. Groundwater is the dominant source of irrigation in the rain-fed regions and plays a key role in the socio-economic development of agrarian economies. Since ground-water remains hidden in a complex system of rocks, its precise assessment is difficult, and this has resulted in a large mismatch between groundwater demand and availability. While watershed interventions are expected to improve groundwater recharge through better soil and water conservation practices, the actual availability of groundwater for final use depends on the suitability of interventions to the aquifer system. Chapter 4 provides a comprehensive assessment of the interventions at the household level and also discusses the appropriateness of the interventions to the respective hydrological regime.

The need for sustainable groundwater management cannot be over emphasized, especially in hard rock harsh environments like Anantapuramu district. Sustaining groundwater resources and ensuring equitable access to the resource is crucial for enhancing the resilience as well as sustaining the livelihoods. While groundwater collectivization and social regulation are expected to address these issues, the effectiveness of such initiatives needs to be assessed arid and semi-arid conditions where groundwater resources are extremely scarce. In Chap. 5, the institutional modalities of groundwater collectivization and their effectiveness (in terms of sustainability and equity) in arid and semi-arid regions of Anantapuramu district are presented. While the AF-EC initiatives of groundwater collectivization in recent years represent the arid conditions, some of the earlier interventions by other NGOs in the district have been in the semi-arid region. A comparative assessment of these two regions helps to understand the potential and constraints of participatory groundwater institutions in harsh environments for scaling up in future.

The lessons drawn from the comprehensive and integrated approach towards mitigating climate/drought risks provide useful insights for scaling up initiatives across similar agroecological contexts across India and elsewhere. The constraints associated with the existing approaches and policy support that is needed to overcome them need to be identified for making the interventions more effective. Based on the experiences of AF-EC in addressing climate/drought risks, Chap. 6 draws lessons and policy imperatives for improving the welfare of the communities in the region.

Chapter 2
Socio-ecological Profile of the Sample Watersheds

2.1 Background

Ananthapuramu district, which was formed in 1882, lies between 13′–40′ and 15′–15′ northern latitude and 76′–50′ and 78′–30′ eastern longitude. Total geographical area of the district is 19.13 lakh hectares. The district is divided into five revenue divisions (Ananthapuramu, Dharmavaram, Penukonda, Kadiri, Kalyandurgam) consisting of 63 revenue Mandals and 964 revenue villages. As per 2011 census, the population of Ananthapuramu was 0.41 million, and the share of rural is 72% and of urban is 28% with an average density of 213 per sq. km. and sex ratio of 977. About 20% of the population comprises scheduled caste (SC) and scheduled tribe (ST), who belong to the lowest rung of the social classification in India. Backward caste (BCs) communities, who are in the middle of the social hierarchy, account for 60% of the population and the remaining belong to other castes (OCs). Literacy rate in the district stands at 64% with 49.9% of its population in work force (see appendix Table 2.1A for district level details). In most demographic indicators, the district is below state average.

Located in southern Andhra Pradesh (AP) in South India, Ananthapuramu is arid and semi-arid and receives the least rainfall of 553 mm average in South India. Southwest monsoon (June–September) accounts for 61% of the total rainfall, and north-east monsoon (October–December) accounts for 39%. It is located at the heart of the "Rain Shadow Region" in Deccan Peninsula. The district has three types of soils, viz. black cotton soils in the north, poor red soils in the centre, and sandy red soils in the remaining part. Overall red soils account for 76% and black soils for 24%. It has thin and scanty forest cover due to its arid conditions – driest part of the state. One of the poorest districts in the country, Ananthapuramu's farmers is largely dependent on chronically drought-prone, rain-fed agriculture. It had seen only two or three normal crops out of 20 years from 1999 to 2019. Approximately 1million ha are cultivated under the drought-prone rain-fed conditions. Mostly a single monocrop of groundnut is sown in about 0.8 million ha under such harsh and

© Springer Nature Switzerland AG 2020
V. R. Reddy et al., *Climate-Drought Resilience in Extreme Environments*,
https://doi.org/10.1007/978-3-030-45889-8_2

agroclimatic conditions. Only about 0.1 million ha (10% of net cropped area) is under irrigation, that too mostly under undependable tube wells and local surface water bodies which are dependent again on the local rain. There is severe pressure on groundwater, and the incidence bore wells drying up is common. Majority of the farmers own less than 2 ha of land, i.e. 93% of the farmers (0.56 million out of 0.6 million farmers) are small and marginal.

The district is mainly agriculture-dependent with net sown area of 0.84 million hectare with a crop intensity of 112%. The district occupies the lowest position in respect of irrigation facilities with only 16% of the gross cropped area (GCA) being irrigated during 2014–15. During the 1970s the cropping pattern in the district had shifted away from millets in favour of groundnut, viz. the share of area under groundnut was between 60% and 75% during 1980s and 1990s. Of late, there is again a shift towards multiple cropping and reduction in area under groundnut and paddy. Agriculture in the district is characterized with uncertainty of rainfall coupled with limited irrigation sources affecting crop area and yields. For instance, over the last 20 years in Ananthapuramu district, the area under groundnuts varied by a factor of two and yield by a factor of 20 (between 1310 and 67 kg/ha). With virtually no other industry, Ananthapuramu's backwardness and poverty are well indicated in its severe rural indebtedness, farmers abandoning agriculture and resorting to wage employment, distress migration to cities, and increasing number of farmer suicides. Malnutrition, illiteracy, illnesses, deprivation, and caste and gender discrimination are higher than state average.

Given the harsh environmental conditions along with poor soil conditions and water scarcity, Watershed development (WSD) is being implemented in the district on a priority basis over the last three decades. Over the years the programme is being adopted to changing climatic conditions and increasing vulnerabilities of the communities. Sample watersheds were implemented with specific emphasis on drought resilience and livelihoods. The aim is to develop and strengthen socio-economic conditions of the people through conservation of resources by implementing appropriate physical and biological interventions. The developmental activities that are required under the project are planned by the stakeholders and implemented by the community. Awareness-building activities regarding environmental degradation and the need for its prevention and restoration, augmentation of water resources, improving vegetative cover, and alternative livelihoods through acquisition of better skills and knowledge are being carried out. Physical interventions to conserve soil, to improve moisture status, to harvest rain water to improve groundwater recharge, and to improve land cover by promoting horticulture and planting trees in common lands are undertaken. These programmes are expected to change/improve the mind-set of the people and make them active partners in the development. In what follows, a brief profile of the sample watersheds is presented .

2.2 Profile of the Selected Watersheds

IWMP (Muttala) watershed is located in Atmakur Mandal of Ananthapuramu district. The demarcated watershed is divided into four micro-watersheds spread over four villages. Area covered (treated) by watershed ranges from 1100 to 3000 acres (Table 2.1). The NABARD watershed is located in the Kalyandurg Mandal with watershed treated area of 2062 acres and a population of 1084. In terms of households the size of watershed villages ranges between 148 and 429 households, i.e. neither too big or too small. The control village (IWMP) Kurlapalli is the smallest of the sample villages with 78 households with an area of 750 acres, while the NABARD control village Narayanapuram (N. Puram) is the largest with 428 households. In majority of the villages, backward caste/community (BC) is dominant, though one of the villages has 98% of SC population. There are no OC households in Devankunta Thanda (D. K. Thanda) and Kurlapalli. The average rainfall is about 352 mm. There is a significant change in the amount of rainfall and also potential evapotranspiration observed in these areas. These changes have drastic effect on crop production activities in the region and have increased the risks of farming because of change in length of growing period.

Biophysical characteristics of the sample watersheds vary. The average rainfall is around 540 mm for NABARD watershed, while it is as low as 341 for the IWMP watershed villages (Table 2.2). Proportion of area under forest is nil in the NABARD watershed village, while it ranges between 13% and 30% in the IWMP villages. Variations in precipitation and forest coverage are reflected in moisture availability and groundwater levels in the watershed villages.

Table 2.1 Details of the sample watersheds

Watershed	Geographical/ treated area (acres)	# Households (population: M/F)	% of SC/ST population	% of BC population	% of OC population
D. K. Thanda	1520/1133.5	148 (396/375)	98 (only SC)	02	0
Goridindla	1876.3/1690.7	266 (524/413)	12	73	15
Muttala	3276/3066	429 (854/714)	08	65	27
Papampalli	1528/1320	172 (338/326)	33	18	49
IWMP Total	*8200.3/7210.2*	*1015 (2112/1828)*	*29*	*48*	*23*
Battuvanipalli (NABARD)	2812.5/2062.5	214 (556/528)	32	39	29
Kurlapalli (C-IWMP)	875/750	78 (275/260)	05	95	0
N.Puram (C-NABARD)	2850/2400	428 (650/720)	25	21	54

Source: Village Records
Note: #, Number of; C-IWMP, control village of IWMP watershed; C-NABARD, control village of NABARD watershed. M/F, male/female; D. K. Thanda, Devankunta Thanda; N. Puram, Narayanapuram

Table 2.2 Biophysical details of the sample watersheds

Watershed	Average rainfall (mm)	% Forest area	GWL (mtrs)	Available soil moisture till 30 cm depth (%)
D. K. Thanda	341	13	15.7	74
Goridindla	341	13	15.7	74
Muttala	341	30	15.7	74
Papampalli	341	30	19.7	74
IWMP Total	*341*	*17*	*18*	*74*
Battuvanipalli (NABARD)	542	0	38.7	65
Kurlapalli (C-IWMP)	341	14	15.7	74
N.Puram (C-NABARD)	542	0	26.6	65

Source: Village Records
Note: #, number of; C-IWMP, control village of IWMP watershed; C-NABARD, control village of NABARD watershed. M/F, male/female; D. K. Thanda, Devankunta Thanda; N. Puram, Narayanapuram; GWL, groundwater level in meters below ground

Distribution of sample households is close to population distribution across the sample villages, as proportionate sampling by landholding size (including landless) was drawn (Table 2.3). Large farmers (> 10 acres) are mostly OC farmers, while landless (0 acres) and small and marginal (< 5 acres) farmers are concentrated among SC households. Even in the complete SC village (D. K. Thanda), only 10% of the SC households are large farmers. Among OC households the share of large farmers range between 14% and 68% among the watershed villages, while only 6% to 18% among the BC households. That is, land concentration continues to be with socially higher up communities. Average farm size ranges between 6.3 and 10.6 acres among the watershed villages indicating skewed distribution of land in favour of large farmers (Table 2.3). Average farm size is substantially higher (10.6 acres) in the NABARD watershed. Among the IWMP watersheds, farm size increases as one moves from upstream to downstream, which is unusual. Compared to control villages watershed villages have larger average holdings.

Importantly, access to irrigation is also more among OC households (Table 2.4). Proportion of area irrigated ranges between 16% and 41% among the watershed villages. In the case of IWMP, watershed villages have substantially higher proportion of area under irrigation when compared to the control village. Whereas, the reverse is true in the case of NABARD watershed. Across the size classes there is no clear bias in favour of large farmers in the distribution of irrigated area. However, OC households have disproportionately higher irrigated area when compared to other communities (Table 2.5).

Average household income is higher among watershed villages when compared to control villages (Table 2.6). NABARD watershed has higher household income when compared to IWMP watershed villages. OC households have higher income when compared to other communities. In general, medium and large farmers have higher household income due to their greater access to land and water. In some

Table 2.3 Average farm size of Sample Households by socio-economic groups (in acres)

Watershed	SC/ST				BC				OC				All
	S&M	Medi.	Large	SC/ST	S&M	Medi.	Large	BC	S&M	Medi.	Large	OC	
D. K. Thanda (US)	4.3	6.8	11.9	6.3	0	5.5	0	5.5	0	0	0	0	6.3
Goridindla (MS)	5	6	0	5.4	4.7	6.7	11.3	6.1	4.5	7.7	15	7.8	6.3
Muttala (MS)	3.5	8.5	0	5.2	3.9	7.9	16.3	8.1	3.8	8.4	16.7	9.5	8.3
Papampalli (DS)	2.6	8.1	0	4.6	3.5	7	0	6	4.2	7.6	20.6	13.5	9.2
Battuvanipalli	4.1	0	0	4.1	3.6	7.8	11	5	5	9.5	21.5	17.9	10.6
Kurlapalli (C-IWMP)	5	0	0	5	3.6	7.9	14.8	6.5	0	0	0	0	6.4
N.Puram (C-NABARD)	2.6	0	0	2.6	3.5	8.1	14	5.9	3	7.7	11.5	8.3	5.4

Source: Field Survey

Note: S&M, small and marginal; Medi, medium. SC/ST, scheduled caste/scheduled tribe; BC, backward castes; OC, other castes

Table 2.4 Proportion of irrigated area by sample household by socio-economic groups (% area)

Watershed	SC/ST				BC				OC				All
	S&M	Medi	Large	SC/ST	S&M	Medi	Large	BC	S&M	Medi	Large	OC	
D. K. Thanda	35	24	47	32	0	9	0	9	0	0.0	0	0	32
Goridindla	33	42	0	37	37	24	15	30	100	76	67	83	38
Muttala	33	0	0	22	50	44	29	43	32	20	83	42	41
Papampalli	10	11	0	11	0	19	0	13	8	32	24	21	16
Battuvanipalli	3	0	0	3	14	41	36	21	40	15	33	30	21
Kurlapalli (C-IWMP)	0	0	0	0	8	14	24	12	0	0	0	0	12
N.Puram (C-NABARD)	16	0	0	16	49	24	28	40	0	49	50	39	32

Source: Field Survey

Note: S&M = small and marginal; Medi; medium. SC/ST, scheduled caste/scheduled tribe; BC, backward castes; OC, other castes

Table 2.5 Distribution of irrigated area of sample households by socio-economic groups (% area)

Watershed	SC/ST				BC				OC			
	S&M	Medi.	Large	SC/ST	S&M	Medi.	Large	BC	S&M	Medi.	Large	OC
D. K. Thanda	34	31	36	99	0	100	0	1	0	0	0	0
Goridindla	50	50	0	10	51	39	10	51	24	49	27	38
Muttala	100	0	0	1	26	40	34	63	7	14	80	36
Papampalli	41	59	0	8	0	100	0	15	3	14	82	77
Battuvanipalli	100	0	0	1	33	48	19	17	2	6	92	82
Kurlapalli (C-IWMP)	0	0	0	0	18	39	43	100	0	0	0	0
N.Puram (C-NABARD)	100	0	0	4	53	25	21	51	0	34	66	45

Source: Field Survey

Note: S&M, small and marginal; Medi, medium; SC/ST, scheduled caste/scheduled tribe; BC, backward castes; OC, other castes

Table 2.6 Average household income of sample households by socio-economic groups (in Rs. Lakhs/HH)

Watershed	SC/ST					BC					OC					All
	LL	S&M	Medi	Large	SC/ST	LL	S&M	Medi	Large	BC	LL	S&M	Medi	Large	OC	
D. K. Thanda	0.6	0.88	0.91	1.47	0.90	0	0	0.90	0	0.90	0	0	0	0	0	0.9
Goridindla	1.1	0.8	1.4	0	1.0	0.6	1.0	1.0	1.4	0.9	0.5	1.0	2.5	4.0	2.0	1.1
Muttala	0	0.9	3.0	0	1.6	0.5	0.9	1.5	1.4	1.1	1.1	0.9	1.2	3.2	1.6	1.3
Papampalli	1.2	0.8	1.2	0	1.0	0.9	0.5	0.9	0	0.8	0.4	1.0	1.4	2.7	1.8	1.3
Battuvanipalli	0.5	1.4	0	0	1.0	0.9	0.7	1.5	1.8	1.0	0.9	1.0	1.4	2.5	2.1	1.4
Kurlapalli (C-IWMP)	0.2	0.3	0	0	0.6	0.7	0.5	1.2	1.5	0.9	0	0	0	0	0	0.8
N.Puram (C-NABARD)	0.3	0.9	0	0	0.8	0.5	1.0	1.9	1.5	1.1	0.9	0.5	1.1	3.8	1.8	1.2

Source: Field Survey
Note: LL, landless; S&M, small and marginal; Medi, medium; SC/ST, scheduled caste/scheduled tribe; BC, backward castes; OC, other castes

cases, landless have higher household income when compared to small and marginal farmers. This indicates that income from nonfarm activities contributes substantially in some villages.

2.3 Hydrogeology of Watershed Sites

As indicated earlier groundwater is the major source of water in the region. Understanding the hydrology of the watershed sites is critical for assessing the potential and suitability of locations for groundwater recharge. Watershed interventions like constructing recharge structures in terms of their location and density depend on the geometry of the aquifer. For this purpose, thematic layers are used in Geographic Information Systems (GIS) environment. Remote sensing data which comprises of 30 m interval Shuttle Radar Topographic Mission (SRTM) and digital elevation model (DEM) and geology, structures, and groundwater prospect maps along with Survey of India (SoI) toposheets and field observation data were used. Based on drainage and elevation, micro-watersheds are divided in to upstream, midstream, and downstream. D. K. Thanda (Devankunta) comes under upstream, Goridindla and Muttala are located in midstream, and Papampalli comes under downstream (Fig. 2.1). Based on the hydro-geomorphological units, the study area is classified into four zones (Figs. 2.2 and 2.3). Recharge zones are categorized as good, moderate, very low, and run-off zones. Good recharge zones are suitable for artificial recharge structures like check dams, *nala* bunds and percolation tanks. Moderate recharge zones are suitable for check dams, and very low recharge zones are suitable for recharge pits. Run-off zone requires no artificial recharge interventions. It may be noted that the sample IWMP (Muttala) watershed is characterized with high proportion of very low recharge zone areas followed by medium, high, and run-off regions (Fig. 2.2). As one moves from high elevation (upstream/ridge) to low elevation (downstream/valley), very low recharge zone increases. Upstream locations are of moderate recharge zone interspersed with high recharge zones. High recharge zones are also available in the downstream locations as well. Run-off zone is mostly concentrated on the periphery of downstream locations.

In the case of NBARD (Buttuvanipalli) watershed, the aquifer geometry indicates that good recharge zone is located in the central part of watershed, while the run-off and very low recharge zones are located at the periphery (Fig. 2.3). This clearly indicates that groundwater recharge potential will not necessarily be higher in the downstream locations. In both the watersheds, high and moderate recharge zones are available in the upstream locations. In the case of IWMP (Muttala) watershed, very low recharge zone is located only in the downstream locations. The nature and density of interventions need to be in tune with the recharge zone in order to optimize the benefits.

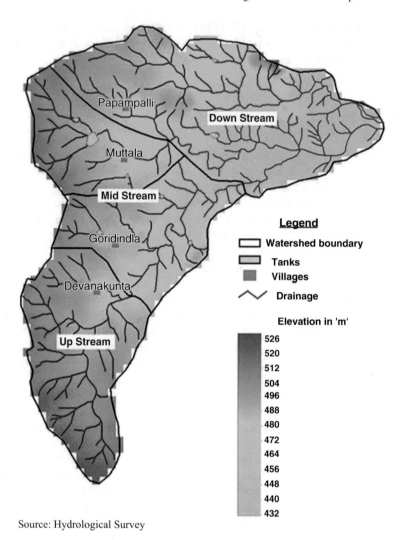

Source: Hydrological Survey

Fig. 2.1 Basic features of IWMP Watershed Project Muttala. (*Source*: Hydrological Survey)

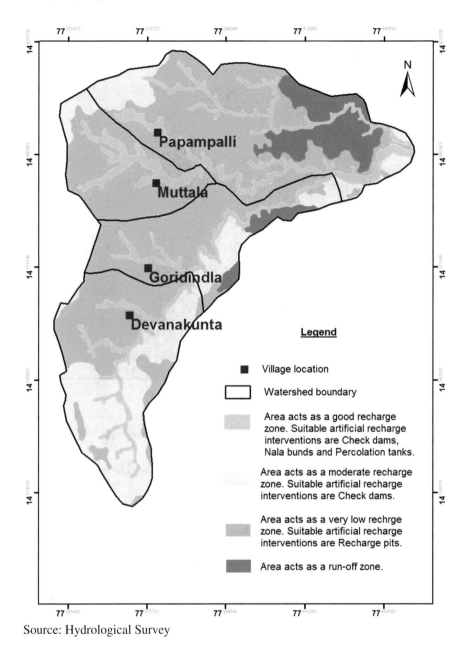

Source: Hydrological Survey

Fig. 2.2 Areas suitable for artificial recharge interventions – IWMP Watershed Project Muttala. (*Source*: Hydrological Survey)

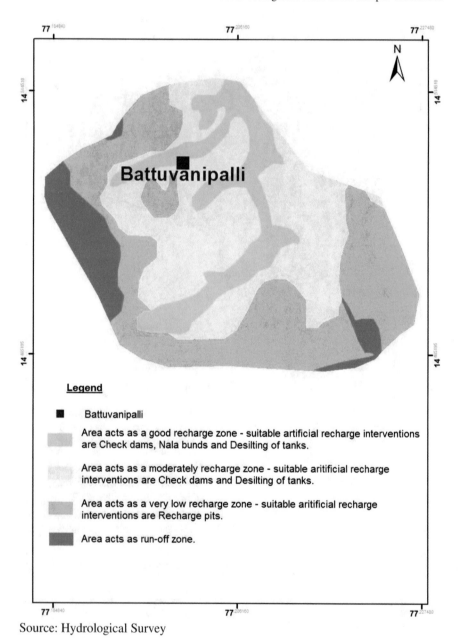

Source: Hydrological Survey

Fig. 2.3 Area suitable for artificial recharge interventions – NABARD Watershed Battuvanipalli. (*Source*: Hydrological Survey)

Appendix

Table 2.1A Comparison of the district with the state

Sl. No.	Item	Unit	Year	State	District
_1_	_Geographical area:_	000'sq. kms.	2011	163.00	19
_2_	_Population as per 2011 census_				
(a)	Total population	In lakhs	2011 census	495.77	40.81
(b)	Male population	In lakhs	2011 census	248.30	20.64
(c)	Female population	In lakhs	2011 census	247.47	20.17
(d)	Sex ratio	Females for 1000 males	2011 census	997	977
(e)	Rural population	In lakhs	2011 census	349.67	29.35
(f)	% of rural population to total population	In lakhs	2011 census	70.53	71.93
(g)	Urban population	In lakhs	2011 census	146.1	11.46
(h)	% of urban population to total population	Percentage	2011 census	29.47	28.07
(i)	No. of households	In lakhs	2011 census	127.19	9.68
(j)	Density of population	Persons per sq. km.	2011 census	304	213
(h)	Growth rate over the previous census (2001-2011)	Percentage	2011 census	9.21	12.10
_3_	_Child population (0-6)_				
(a)	Total population	In lakhs	2011 census	52.44	4.46
(b)	Male population	In lakhs	2011 census	26.97	2.31
(c)	Female population	In lakhs	2011 census	25.47	2.15
(d)	Rural population	In lakhs	2011 census	37.83	3.24
(e)	Urban population	In lakhs	2011 census	14.61	1.22
_4_	_Literates_				
(a)	Total literates	In lakhs	2011 census	298.60	23.11
(b)	Male literates	In lakhs	2011 census	165.50	13.38
(c)	Female literates	In lakhs	2011 census	133.10	9.73
_5_	_Literacy rate_				
(a)	Total literacy rate	Percentage	2011 census	67.35	63.57
(b)	Male literacy rate	Percentage	2011 census	74.77	73.02
(c)	Female literacy rate	Percentage	2011 census	59.96	53.97
_6_	_No. of households_				
(a)	Total	In lakhs	2011 census	127.19	9.68
(b)	Rural	In lakhs	2011 census	90.65	7.00
(c)	Urban	In lakhs	2011 census	36.54	2.68
_7_	_Household size_				
(a)	Total	No. of persons/HH	2011 census	4	4
(b)	Rural	No. of persons/HH	2011 census	4	4

(continued)

Table 2.1A (continued)

Sl. No.	Item	Unit	Year	State	District
(c)	Urban	No. of persons/HH	2011 census	4	4
8	*Scheduled caste population*				
(a)	Total population	In lakhs	2011 census	84.69	5.83
(b)	Male population	In lakhs	2011 census	42.20	2.92
(c)	Female population	In Lakhs	2011 census	42.49	2.91
(d)	% of SC population to total population	Percentage	2011 census	17.08	14.29
9	*Scheduled tribe population*				
(a)	Total population	In lakhs	2011 census	27.40	1.54
(b)	Male population	In lakhs	2011 census	13.62	0.78
(c)	Female population	In lakhs	2011 census	13.78	0.76
(d)	% ST population to total population	Percentage	2011 census	5.53	3.78
10	*Workers*				
(a)	Total workers (main+marginal)	In lakhs	2011 census	230.81	20.36
(b)	Agriculture workers (cultivators+agrl. labour)	In lakhs	2011 census	143.93	10.26
(c)	Non-agriculture workers (Household industries+others)	In lakhs	2011 census	86.88	6.53
11	*Crude birth rate (United A.P.)*				
(a)	Total	Rate	S.R.S -2013	17.4	16.90
(b)	Rural	Rate	S.R.S -2013	17.7	16.40
(c)	Urban	Rate	S.R.S -2013	16.7	17.40
12	*Crude death rate (United A.P.)*				
(a)	Total	Rate	S.R.S -2013	7.3	7.1
(b)	Rural	Rate	S.R.S -2013	8.3	8.20
(c)	Urban	Rate	S.R.S -2013	5	6.00
13	*Infant mortality rate (United A.P.)*				
(a)	Total	Rate (%)	S.R.S -2013	39	45
(b)	Rural	Rate (%)	S.R.S -2013	44	46
(c)	Urban	Rate (%)	S.R.S -2013	29	44
14	*Villages/Gram Panchayats/Mandals/Mandal Parishads)*				
(a)	Census villages (including uninhabited villages)	Nos.	Census 2011	17,366	949
(b)	Revenue villages	Nos.	2016	17,751	964
(c)	Gram Panchayats	Nos.	2016	12,918	1003
(d)	Revenue Mandals	Nos.	2016	670	63
(e)	Mandal Parishads	Nos.	2016	654	63
(f)	Revenue divisions	Nos.	2016	49	5
(g)	Towns (Statutory and Census, 2011)	Nos.	Census, 2011	195	7
(h)	Municipal corporations	Nos.	2016	13	1
(i)	Municipalities/Nagar Panchayats	Nos.	2016	97	11

(continued)

Table 2.1A (continued)

Sl. No.	Item	Unit	Year	State	District
15	*Rainfall*				
(a)	Normal rainfall	In mms	2015–2016	912.5	553.00
(b)	Actual rainfall	In mms	2015–2016	966	608.00
16	*Agriculture*				
(a)	Gross cropped area	In '000' Hect.	2015–2016	7,532	921.00
(b)	Net cropped area (including fish ponds)	In '000' Hect.	2015–2016	16,169	849.00
(c)	Gross irrigated area	In '000' hect.	2015–2016	3,547	174.00
(d)	Net area irrigated	In '000' hect.	2015–2016	2,743	171.00
	(A)Area under				
(a)	Rice	In '000' hect.	2015–2016	2,161	33.59
(b)	Jowar	In '000' hect.	2015–2016	175	30.84
(c)	Maize	In '000' hect.	2015–2016	233	17.84
(d)	Red gram	In '000' hect.	2015–2016	220	38.51
(e)	Green gram	In '000' hect.	2015–2016	211	27.54
(f)	Black gram	In '000' hect.	2015–2016	456	0.78
(g)	Bengal gram	In '000' hect.	2015–2016	471	75.79
(h)	Groundnut	In '000' hect.	2015–2016	775	468.18
(i)	Sunflower	In '000' hect.	2015–2016	26	5.33
(j)	Chillies	In '000' hect.	2015–2016	156	3.79
(k)	Onion	In '000' hect.	2015–2016	41	3.12
(l)	Sugarcane (gur)	In '000' hect.	2015–2016	122	0.16
(m)	Cotton	In '000' hect.	2015–2016	666	60.31
(n)	Tobacco	In '000' hect.	2015–2016	98	0.21
	(B) Production				
(a)	Rice	'000 tonnes	2015–2016	11,233	4,356
(b)	Jowar	'000 tonnes	2015–2016	358	486
(c)	Maize	'000 tonnes	2015–2016	1,412	3,120
(d)	Red gram	'000 tonnes	2015–2016	132	216
(e)	Green gram	'000 tonnes	2015–2016	137	149
(f)	Black gram	'000 tonnes	2015–2016	411	803
(g)	Bengal gram	'000 tonnes	2015–2016	500	451
(h)	Groundnut	'000 tonnes	2015–2016	801	752
(i)	Sunflower	'000 tonnes	2015–2016	23	546
(j)	Chillies	'000 tonnes	2015–2016	618	3,041
(k)	Onion	'000 tonnes	2015–2016	696	720
(l)	Sugarcane (gur)	'000 tonnes	2015–2016	937	2,009
(m)	Cotton Lint	Bales of 170 Kgs. each	2015–2016	1,817	1,198
(n)	Tobacco	'000 tonnes	2015–2016	222	338

(continued)

Table 2.1A　(continued)

Sl. No.	Item	Unit	Year	State	District
17	*Veterinary facilities*				
	Total livestock population	In Lakhs	As per 2012 census(P)	294.03	126.40
(a)	Super Speciality Veterinary Hospital	Nos.	2015–2016	2	
(b)	Veterinary polyclinics	Nos.	2015–2016	12	1
(c)	Veterinary hospitals	Nos.	2015–2016	180	16
(d)	Veterinary dispensaries	Nos.	2015–2016	1420	110
(e)	Rural livestock units including mobile veterinary clinics	Nos.	2015–2016	1505	64
(f)	Artificial insemination centres	Nos.	2015–2016	6029	175
18	*Hospitals*				
(a)	Govt. hospitals (allopathic)(general, allied, CHCs)	Nos.	2015–2016	260	116
(b)	Hospitals for special treatment	Nos.	2015–2016	20	
(c)	Primary health centres	Nos.	2015–2016	1,075	80
(d)	Dispensaries including ESI diagnostic centres	Nos.	2015–2016	81	
(e)	Hospitals and dispensaries (Ayurvedic)	Nos.	2015–2016	628	22
(f)	Hospitals and dispensaries (Homoeopathy)	Nos.	2015–2016	361	25
(g)	Hospitals and dispensaries (Unani)	Nos.	2015–2016	161	19
(h)	Hospitals and dispensaries (Naturopathy)	Nos.	2015–2016	44	2
(i)	No .of doctors in all hospitals (including contract)	Nos.	2015–2016	5,620	350
(j)	No. of beds in all govt. hospitals	Nos.	2015–2016	28753	1894.0
19	*Villages electrified*				
(a)	No. of inhabited villages	Nos.	2011 census	16837	929
(b)	Percentage of villages electrified	%		100	100
20	*Education*				
	(A) No. of institutions:				
(a)	Degree colleges (govt.+private aided)	Nos.	2015–2016	271	64
(b)	Junior colleges	Nos.	2015–2016	3,315	220
(c)	Schools including elementary, UPS, high schools, and higher secondary schools	Nos.	2015–2016	61,128	5021
(d)	B.Ed. training colleges (govt.+private)	Nos.	2015–2016	378	25
(e)	Polytechnic (govt.+private)	Nos.	2015–2016	317	15
(f)	Engineering colleges (govt.+private)	Nos.	2015–2016	368	17
(g)	Medical colleges(including private Colleges) (Allopathy) (MBBS)	Nos.	2015–2016	25	1

(continued)

Table 2.1A (continued)

Sl. No.	Item	Unit	Year	State	District
(h)	Pharmacy (govt.+private)	Nos.	2015–2016	128	3
(i)	I.T.I (govt.+private)	Nos.	2015–2016	477	36
(j)	M.B.A. (govt.+private)	Nos.	2015–2016	396	23
(k)	M.C.A. (govt.+private)	Nos.	2015–2016	194	16
	(B)Students enrolled in colleges				
(a)	Degree colleges (govt.+private aided) (excluding oriental)	Nos.	2015–2016	217238	54224
(b)	Junior colleges	Nos.	2015–2016	555319	56356
(c)	Schools including elementary, UPS, high schools, and higher secondary schools	Nos.	2015–2016	7222771	580403

Source: Chief Planning Officer (2018), Hand Book of Statistics – Ananthapuramu district – 2018, Ananthapuramu

Chapter 3
Implementation Process: Quality, Equity, and Sustenance

3.1 Introduction

Proper implementation of any programme is the key to its success. Following the programme guidelines does not necessarily guarantee better implementation. Quality and sustenance of interventions in the long run depends on not only the institutional capacities of the project implementing agency (PIA) but also on the systems it adopts and institutions it promotes and strengthens at the village level. Besides, communities need to be capacitated to understand and supervise various activities under the programme. AF-EC has more than 30 years of experience in implementing developmental programmes including WSD. AF-EC adopts an integrated approach of science to development as well as practice to policy. It adopts a multipronged approach of integrating technologies, sustainable resource management, farming systems, and practices that support and sustain livelihoods. Participatory approach is at the core of its developmental interventions. People's participation is ensured from planning to monitoring and evolution. More importantly, equal participation of women and other vulnerable sections of the community is ensured and promoted for inclusive livelihood benefits. The AF-EC not only works with scientific community and organizations like agriculture universities in developing appropriate technologies and crop systems for the region but also improvises on the existing technologies and techniques in order to suit the changing agroclimatic conditions in the district. These technologies, techniques, and farm practices are demonstrated for the benefit of farmers for wider adaptation and location-specific modifications. Effective techniques and practices are showcased to other stakeholders like policymakers and development practitioners for scaling up and scaling out.

The main focus is human resources and institutional evolution and strengthening. AF-EC has committed and trained staff. The staff is well trained in technical and social aspects of developmental interventions such as people's participation, consensus building, conflict resolution, leadership, and community-based

© Springer Nature Switzerland AG 2020
V. R. Reddy et al., *Climate-Drought Resilience in Extreme Environments*,
https://doi.org/10.1007/978-3-030-45889-8_3

institutional development. Emphasis is given to capacitate communities in terms of watershed activities as well as alternative livelihood capabilities. Their commitment and innovative approaches have been recognized at the state and central level. AF-EC has been honoured with "best performance award" for implementing the IWMP watersheds in the state by the state government for the years 2009 and 2012. And some of their watersheds are notified as model watersheds for exposure visits to other PIAs and farmers. Though programme-specific guidelines are followed while implementing various programmes, the centre has developed its own approach in identifying the villages and beneficiaries across the programme. At present AF-EC is implementing WSDPs funded by different sources, viz. (i) IWMP funded by Government of India (Ministry of Rural Areas and Employment) through the Department of Rural Development, Government of AP (GoAP), and (ii) watershed development funded (WSDF) under WSDF of NABARD. These programmes have different guidelines and approaches to implementation, though some common practices are adopted by AF-EC in both the cases. Here the implementation process for these two programmes is presented.

3.2 Integrated Watershed Management Programme (IWMP) – Muttala

AF-EC has been implementing three *mega* watersheds, viz. (i) Muttala in Atmakur Mandal, (ii) Bandameedapalli in Rapthadu Mandal, and (iii) Kuderu in Kuderu Mandal. Muttala mega watershed is from the first batch of IWMP watersheds (2009–2010). Its implementation was completed in 2016–2017. Muttala *mega* watershed consists of four micro-watersheds (villages covering) 2535 ha of treated area. The process of implementation was initiated with the awareness campaign articulating the importance of watershed interventions and the need for community's participation in the process. The modalities of the implementation were discussed among different sections of the community, viz. elders, opinion leaders, *Gram Panchayat*[1] (GP) representatives, community-based organizations (CBOs), etc., Once there was consensus for taking up the watershed programme, *Grama Sabha* (GS: village general body) was conducted in each village (micro-watershed), where the rationale and relevance of participatory Watershed Management (WSM) emphasizing the need for sustainable natural resource management (SNRM) at the village level was explained.

The first task of the *Grama Sabha* is to create a watershed development committee (WDC) consisting of 13–15 members. These members should represent all sections of the community (landless, SC/ST; small and marginal and women farmers/households) and selected by consensus. WDC is responsible for implementing the

---

[1] Democratically elected body at the village level. It consists of ward members and a president and responsible for village level administration.

WSD activities as per the detailed project report (DPR) in a transparent process ensuring quality in a given time frame. WDC is headed by a chairperson, elected from among the committee members and by the members. WDC selects one active unemployed educated youth to work as "Watershed Assistant" to supervise the implementation of works and report to WDC & PIA periodically. WDC also undertakes field visits periodically to ensure the quality and timeliness of works. In coordination with the federation of women SHGs, i.e. village organization (VO), WDC selects the beneficiaries from poor families to benefit from the livelihood enhancement component of the IWMP. WDC and VO are responsible for managing the livelihood funds including the recovery of the loans from the beneficiaries. The committee is expected to meet once in 15 days to review the progress of works and pass resolutions to take up new works as per the DPR.

Within each micro-watershed, a number of user groups (UGs) are formed in order to make the participation more effective and cohesive. One UG for every 100–120 acres are formed. Sometimes UGs are formed on the basis of specific activity, viz. dryland horticulture, fodder development, etc. UG members contribute Rs. 50 per month per member; a matching grant is provided under the IWMP from the State Level Nodal Agency (SLNA). These funds are used for buying new equipment, seeds, etc., for the group. In Muttala watershed 44 UGs have been formed.

DPR preparation is a critical first step in the watershed implementation process. It is observed that DPR preparation is often outsourced to other agencies instead of the PIA involving in its preparation (LNRMI 2012). In the present case the PIA (AF-EC) itself prepared the DPR adopting a stepwise approach. Participatory rural appraisal (PRA) exercises were conducted in each micro-watershed (village) involving all sections of the community. The process of PRA exercises helped the communities to know and understand the status of their natural and social resources, viz. drawing of resource and social maps themselves depicting various natural resources and social structures existing in the village. A transect walk was taken up to perambulate the area to assess the land configuration, soil types, existing vegetation, etc. The second step involved conducting of socio-economic survey at the household level with the help of local educated youth. Socio-economic survey was conducted using a structured questionnaire. The survey was followed by a detailed net planning. Net planning relies heavily on consultations with the farmers in their own fields. The type and location of interventions agreed with farmers are marked both on the ground and on land holding maps. Fields (and, often, areas within fields) are assessed for slope, soil depth, soil texture, and erosion status. Net planning helps arriving at clear financial implications though most PIAs don't adopt net planning. Besides, livelihood improvement activities at the village level were also prepared for 5 years.

All these components form the DPR, which is the basis for planning and implementing soil and moisture conservation (SMC) works, rainwater harvesting structures (RWHS), dryland horticulture (DLH), vegetation improvement works, IGAs, etc. Often these activities are planned as per the unit costs, which are fixed at Rs. 12,000 per ha. In the case of AF-EC, the works and activities are planned as per the requirement and any shortfall in IWMP budget is met from the Mahatma Gandhi

National Rural Employment Guarantee Scheme (MGNREGS) funds. The plan considers the IWMP as well as MGNREGS funds to make a total WSD plan. The convergence of IWMP and MGNREGS takes place at the village level due to the broader involvement of the community and the institutions, viz. PRI, SHGs, CBOs, etc., Such a convergence is being planned at the national level in a number of states, but yet to arrive at feasible operational modalities. The AF-EC approach suggests that convergence at the community level with an inclusive approach is critical for programmatic convergence. The policy advocacy role of AF-EC (demonstrating the benefits of convergence) was instrumental in convincing the line departments at the district and state levels and plays a supportive role. That is, a "bottom-up" approach rather than a "top-down" approach is more effective in bringing the convergence.

AF-EC adopted a transparent online payment system (OPS), i.e. electronic fund management system (EFMS). A Watershed Computer Centre (WCC), which is linked to the SLNA, was established for this purpose. This facilitates SLNA to monitor work progress on a regular basis. SLNA makes payments directly into bank accounts of beneficiaries (skilled and unskilled labour or suppliers) after online verification of muster rolls and materials book (M-Book). This system is highly transparent with regard to financial transactions of the programme as anybody can have access to the website to know the implementation of the programme. For instance, in the case of Muttala watershed the entire planned area (2535 ha.) was treated at a cost of Rs. 6.38 crores benefiting 1200 farmers and generating 3.43 lakh person days of employment by 31 May 2016.

As per the IWMP guidelines, funds are provided for nine activity-specific allocations. These include (i) entry point activity (EPA: 4%), (ii) DPR preparation (1%), (iii) administration (10%), (iv) capacity building (5%), (v) natural resource management (NRM) activities (56%), (vi) productivity systems improvement (PSI: 10%), (vii) livelihood activities (9%), (viii) consolidation (3%), and (ix) monitoring and evaluation (M&E: 2%). Core activities of farming get about 66% of the allocations, while 9% of the funds are provided for livelihood activities, which focus on landless households. These allocations are complemented by the contributions from the beneficiary farmers. An amount equal to 5% of the total cost of the works taken up on each farmers land will be contributed by the farmer to the watershed development fund (WDF). This fund will remain with the WDC, which will be used for post-project maintenance of the interventions.

3.2.1 Activities

Entry point activity (EPA) is taken up to build trust and cooperation among the community. Besides, it helps bringing people together in a collective spirit to fulfil a common need/good. Given the long presence and credibility of AF-EC in the region doesn't require an entry point to gain acceptance of the village community. Since, IWMP provides funding for this and an opportunity to fulfil the most commonly felt need of the community. Access to safe drinking water is identified as one of the

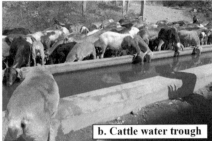

Photo 3.1   Entry point activities (EPAs)

commonly felt needs in the villages, which would benefit all sections. Water purification (RO: reverse osmosis) plants were constructed in every watershed village under the EPA (Photo 3.1a). Along with water purification plants in all the villages cattle, water troughs were constructed to quench the thirst of animals and small ruminants (Photo 3.1b). The EPAs are meant to bring together different sections of people in a village and motivate/encourage them to take decisions in a participatory/ consensus approach. This also helps in better understanding between the PIA and the village community.

IWMP activities could be grouped under NRM, productive system improvement (PSI), livelihoods enhancement (LE), and capacity building (CB) activities. Under the NRM two important components, viz. SMC and RWHS, are taken up. A number of SMC activities (treatment of ridge portions, afforestation in hillocks, treatment of drainage lines, earthen/stone contour bunding, border bunding across the slope, etc.) are taken up. Continuous or staggered contour trenches on ridge portions impound rainwater to improve moisture in upland areas. Block plantations in common lands and planting the upland hilly areas with fodder, fruit and drought hardy biomass species and conserving natural regeneration arrest soil erosion and serve as carbon sinks. Treating the drainage lines with rock-fill dams, gabion structures, loose boulder structures across the first order streams to harvest and reduce the speed of water thus increasing base flows in the drainage system and preventing further deepening of gullies. Stone bunding and new farm bunding (NFB) check soil erosion and improve moisture conservation (Photo. 3.2).

RWH interventions include farm ponds, rock-fill dams (RFDs), check dams (CDs), and percolation tanks (PTs) (Photo 3.3). Farm ponds are constructed with measurements of 8Mt. × 8Mt. × 2Mt. with side slopes of 1:1, in lowest contours of the farmland so as to collect the water drained from upstream side. The water impounded in the farm pond is utilized to provide lifesaving irrigation to rain-fed crops during prolonged dry spells. In some cases, the inner sloping sides of the farm pond and the bottom are to be lined with cement mortar, in order to prevent seepage. In the absence of lining farm ponds recharge groundwater. RFDs and CDs are constructed across streamflows to impound rainwater. While RFDs reduce the velocity of water during heavy downpours by blocking and stopping streamflows, CDs help

Photo 3.2 Soil and moisture conservation (SMC) measures

Photo 3.3 Rainwater harvesting structures (RWHS)

storing the water for 5–6 months. CDs are very useful to increase groundwater and rejuvenate the base flows in streams. The borewells in the vicinity of CD are rejuvenated to give assured irrigation to crops. Check dams are also providing water for cattle, wild animals, etc.

PTs harvest the rainwater over a large area for the purpose of groundwater recharge. RWHS are more popular among the communities, as they provide some

Photo 3.4 Fodder plot and dryland horticulture (DLH)

tangible benefits in short run (within the season). However, the appropriateness of these structures depends on the hydrogeology of the location. In the event of a mismatch between these two, the effectiveness of these structures would be limited.

Dryland horticulture (DLH) and raising fodder plots are specific activities taken up in drought prone areas like Ananthapuramu district (Photo 3.4). DLH is taken up on a large scale in recent years under the IWMP. DLH is not only provides assured income in the medium to long run, it also provides vegetative cover throughout the year apart from facilitating mixed cropping. The most popular horticulture crop in the sample watersheds is mango. Fodder plots are raised by farmers having irrigation. Many farmers with irrigation facility grow fodder crop on at least 0.5–1 acres of land in order to feed the milch animals and small ruminants. Besides, fodder development supports alternative livelihood activity of dairy. Farming community also gets support through PSI component, which aids with procuring farm implements like tillers, cultivators, grass cutting machines, sprayers, etc. Funds are earmarked for providing 50% subsidy and loan portion to provide farm implement packages to UGs, in coordination with Andhra Pradesh State Agro Industries Development Corporation Ltd., (AP AGROS). Custom hiring centres are also established under PSI.

Landless and poor are supported through livelihood activity component. This component provides support for taking up nonfarm IGAs like rearing of milch animals, auto transport, tailoring, petty business, etc. (Photo 3.5). Women are encouraged in all the livelihood activities, including driving. AF-EC runs an independent driving school and provides training for women as well. Tailoring and garment making are also supported in a big way. Beneficiaries receive loan from a revolving/livelihood fund (RF/LF) and repay the same in equal monthly instalments with a nominal rate of interest. A number of programmes that support livelihood activities run in tandem in the villages. Appropriate institutional arrangements are evolved and supported to sustain the livelihood activities in the long run, which is discussed separately.

Photo 3.5 Livelihood activities (LAs)

3.3 NABARD Watershed Development Programme – Battuvanipalli

For its funded watersheds, NABARD adopts a different set of guidelines, which are based on the experience from the IGWSDP under its support. However, most of the AF-EC approaches and practices in watershed implementation are close to these guidelines. The main difference between IWMP and NABARD guidelines is that there is a clear distinction between the capacity building phase (CBP) and the full implementation phase (FIP). During the CBP, village communities and PIAs prepare, plan, implement, and supervise WSD activities. It enables village communities and PIAs to acquire the necessary skills and competency and qualify for inclusion in the FIP. CBP is used to demonstrate different NRM activities on 50 to 100 ha of the watershed area. This provides hands-on experience to the communities. Enough resources and time are allocated for this phase. While the IWMP allocates little time (less than 6 months in practice), NABARD provides 12–18 months for motivating and organizing village communities.

Given the ample time coupled with the demonstration component during CBP, communities are prepared for participation with commitment and conviction. Communities are capacitated in technical skills in soil and water management using the "ridge to valley" concept. During the CBP communities are prepared for undertaking WSDP through "learning by doing" approach. Awareness on WSD was created through village campaigns, *Gram Sabah's*, formal and informal meetings, PRA exercises, and actual implementation of WSD activities on 100 ha. area at the ridge portion in the first 1 year. All appropriate WSD activities would be planned in the 100 ha. of land and implemented in a participatory approach including mobilization of the mandatory *shramadan* (voluntary labour contribution) by the village community.

After satisfactory completion of the CBP for 1 year the programme moves to the full implementation phase (FIP). When the FIP begins a *Grama Sabha* would be conducted to select on consensus 13–15 members from among villagers to form a village watershed development committee (VWDC). The VWDC should have equitable representation for SC & ST communities, small and marginal farmers, landless poor, women, and other weaker sections. The implementation phase begins with a tripartite memorandum of understanding (MoU) between VWDC, PIA, and NABARD. Participatory net planning is then carried out to prepare a 4-year DPR. The DPR would be implemented by the VWDC together with PIA in a transparent and participatory manner. The progress would be regularly monitored by a third-party organization, i.e. Foundation for Ecological Security (FES) besides NABARD.

The entire process is not very different from IWMP and NABARD watersheds, except for creating actual demonstration plot during CBP phase. AF-EC has been following similar approach over the years for implementing WSD. While IWMP was implemented in a cluster of villages or micro-watersheds, NABARD watersheds were limited to one village. Each IWMP micro-watershed is similar to NABARD watershed, similar in size and processes. The formation, roles, and responsibilities of the WDC are similar in both the cases except that there were three watershed supervisors in the case of NABARD instead on one watershed secretary in the case of IWMP. Common interest groups (36) were formed with 5 members in each group. The supervisors employed by VWDC were responsible for three different activities, viz. (i) drainage line treatment, (ii) area treatment, and (iii) livelihood activities. They also assist the technical team in procurement of labour, materials, and measurement of works and ensure quality. They are paid on the basis of their physical and financial performance. Regular trainings are imparted to supervisors by the PIA in order to build their capacity in implementation.

The implementation of Battuvanipalle watershed was completed in December 2015. The project covered 2062.5 acres benefiting 275 households at a cost of Rs. 1.37 crores. Watershed interventions are similar to IWMP, Muttala watershed, which included SMC, RWH, and livelihood activities. Here also water purification plant was set up under EPA. Besides, an office for mutually aided cooperative society (MACS) was constructed under the EPA.AF-EC also facilitated the convergence with other government departments/programmes and organizations. A number of farmers from Battuvanipalli watershed were provided with solar pump sets and drip irrigation systems by the rural development trust (RDT). Many farmers have excavated farm ponds under MGNREGS in their farmlands for critical irrigation purpose. In the case of NABARD, there is wait for convergence funds to come in, as all the interventions are fully funded. Hence, the convergence funds from other programmes like MGNREGS are additional.

Project funding is released to the joint account of WSC and AF-EC (PIA/PFA) in which the money could be withdrawn only with the consent of WSC. The livelihoods component was released as grant to the WSC, and WSC should revolve the amount as loans to the community of the watershed villages, and after the formation and registration of MACS institution, it should be resolved through MACS. The records and the accounts of the project were properly maintained in the WSC office at Battuvanipalli, which indicates transparency in formulation, implementation, execution, and operation of the project. Wage payments were made only after proper scrutiny and inspection of work done by one of the WSC members, and only required amount was withdrawn from the joint account after preparation and verification of the measurement book and muster roll.

Maintenance fund (MF) was created from peoples' contribution, i.e. Rs.100 per year per household (Rs. 45, 470). NABARD also has provided Rs. 3.98 lakh as MF. A grant of Rs.4.83 lakhs was provided by NABARD as village fund. Including interest accrued the total maintenance corpus generated was Rs. 9.57 lakhs by the end of the project. This amount was deposited in a fixed deposit (FD) account of the bank, and only the interest accrued is to be utilized for maintenance of the structures. The interest accrued would be Rs. 0.46 lakhs per annum. In addition to this, the WS community would be contributing to the MF account, as contributions, which is more than sufficient to maintain the project in future. AF-EC would continue to provide the technical expertise for the maintenance of the structures.

3.4 Capacity Building: The Key to Better Implementation and Livelihoods

Development of human capital in a harsh environment like Ananthapuramu is critical for enhancing livelihoods and alleviating poverty. While, there are limits for strengthening the natural capital base, due to geographical and climatic constraints, human capital development has greater potential to expand and alleviate the communities from the dire situation. AF-EC strongly believes in this and has focused and invested in capacity building over the years. AF-EC has an overarching approach to capacity building. All their staff is locally recruited, and most of them come from the programme villages, and many of them have long experience in the organization and have grown within. This automatically ensures better understanding of the local context and also commitment and concern for improving the conditions. Communities also feel comfortable to approach them any time for any kind of support, irrespective of whether there is an ongoing programme in the village or not.

AF-EC has about 60 committed and experienced sociotechnical organizers (STOs) at grass root level. They are all educated rural youth from the district. Many of them are trained and well experienced. Their training and experience include WSC, CBO formation and capacity building, participatory planning and implementation, community organization, etc., Their technical training and experience include a variety of WSD skills like SMC, RWH, horticulture, and sustainable agriculture (SA) including rain-fed agronomical practices, biodiversity, crop diversity, bio-/non-pesticidal management (NPM), bio-fertilizers (organic farming/INM (integrated nutrient management)) like composting, liquid fertilizers, alternate livelihoods for women and youth, etc.

Community level capacity building is also given equal importance. Communities are capacitated not only on WSD but also on crop practices, crop technologies, and livelihood activities. Its approach is not limited to specific programme or intervention/cadre or section of the community. Appropriate training was provided for all the sections of the community, i.e. watershed committee and village organization members, SHG and UG members, landless households, *Sasya Mitra* (friend of plants or trees) and MACS members, etc. (Tables 3.1 and 3.2). Trainings and exposure visits on watershed and farming activities included new agricultural practices, crop systems, input management, horticulture development, maintenance of watershed structures, etc. WSC members were trained on technical aspects of watershed interventions and monitoring system of works. VO members and other farmers were trained on PSI methods. SHG and UG members are provided training on savings and investments, managing loan repayment, etc. Labourers are provided with technical training on farm pond construction and farming. They include alternative livelihoods like garment making, auto driving, managing credit cooperatives, etc.

A result of active and informed participation of the communities had helped in creating high-quality structures. The most common interventions in all the watersheds were check dams, contour trenches, stone bunding, and farm ponds. WSCs are still functioning and look after the maintenance of the structures. In all the villages, we have visited VWC takes care of the structures like check dams, contour trenches, etc., on common lands. VWCs also manage the WSDF created through community contributions. These funds are used as per the requirement of the community.

RFDs/stone outlets, ponds, continuous contour trenches (CCTs)/NFB/trenches, CDs, and horticulture are the important interventions in all the villages (Table 3.3). Interestingly, when compared to earlier watershed interventions in Ananthapuramu District, farm ponds have replaced check dams. One reason could be that in most of the sample villages check dams were already constructed under the earlier watershed interventions. These earlier check dams were repaired under the IWMP in three of the four villages or micro-watersheds. Though watershed interventions were not new to the communities, activities like farm ponds and horticulture are new to them. These activities were introduced in order to enhance the drought resilience of the households. When these interventions are super imposed on the recharge potential (hydrogeology) locations of the watershed, there is a mismatch (see Fig. 3.1). The density of artificial recharge structures is more in the low recharge zones when compared to moderate and high recharge zones. Fine-tuning these interventions is likely to enhance the overall impacts and benefits from the watershed interventions.

Even after 2 years of completion, the watershed structures were maintained well, and most of them are in working condition (Table 3.4). While NABARD watershed has all the structures in working condition in the IWMP watershed, more than 80% of structures are in working condition. None of the structures are non-functional. It may be noted that some of the structures got damaged due to unprecedented heavy

Table 3.1 Details of trainings and exposure visits of IWMP Muttala Project

Training/exposure visit	Participants from	No. of participants	Knowledge gained on
Exposure visits to Tipturu & BAIF NGO's areas of Karnataka state	WDC & VO members	72	Watershed concept interacted the Tipturu & BAIF NGO's WS works and visits to the successful models. Learned the concept of farmer field-level complete utilization of rainwater harvesting activities, every field to be covered by the farm pond, bund plantation, etc.
Capacity building to WS user groups members	Convener & Co-conveners	142	User group concept. How to do savings, NRM work involvement, loan repayment, etc.
Plantation programme for dryland farmers	Horticulture farmers	368	Marking, manure filling, plantation and staking, watering, *maintenance*
Maintenance of NRM works at farmer field level	Farmers	272	NRM activities implementation and *maintenance.*
WDC members and their roles and responsibilities in WSDP	WDC & SHG VO members	72	Roles and responsibilities in organizing and following the regular meetings/ activities and monitoring and supervision of WS works during implementation
Organized a training programme on livelihood activates to the SHG & VO members	SHG VO members	126	Asset creation, beneficiary identification, loan sanction, repayment, follow-up
Project villages tailoring Centre members exposure visits to Pamidi area (garment/handloom industry)	SHG VO members	18	Dress material stretching and marketing issues
Self-employment activity to the educated unemployed youth (auto driving)	SHG members	1	Got trained, issued license, and supported the unit. Started the running of autos from Atmakur to Anantapur and earned of Rs. 400–500 per day
VO & WDC members training on PSI activities	SHG VO & WDC	168	Implementation of PSI, minor irrigation systems, crop diversification, promoting millets, NPM methods, etc.
Farmer training on NRM works	Farmers	132	*Maintenance* NRM assets created through WSDP
Wage seekers execution on farm ponds	Labourers	324	Trained labourers on basic engineering aspects for improvement of skills and abilities on farm pond construction, viz. pit marking, bund sectioning, etc.

Source: PRA/FGDs methods & WS records

Table 3.2 Details of trainings and exposure visits of Battuvanipalli watershed

Capacity Building Programme	No of participants	Participants from	Utility
Training to VWDC	15	VWDC	Strengthening of VWDC
Orientation to the VWDC members on NABARD concept with DGM	3	VWDC	Awareness of watershed programmes and monitoring system
Training to supervisors	4	Supervisors	Orientation to the supervisors on works and records
Productivity enhancement (SRI Paddy)	30	Watershed farmers	To enhance the production of Paddy
Dryland horticulture (DLH) maintenance	30	DLH farmers	Awareness on maintaining of horticulture plots
Exposure visit	15	MACS members	Awareness of MACS

Source: PRA/FGDs methods and watershed records

Table 3.3 Watershed activities taken up in the watershed villages

Activity	D.K. Thanda	Goridindla	Muttala	Papampalle	Battuvanipalle
CCTs/NFB + trench cum bunding	–	1	6	3	12,143 + 30485*
RFDs/stone outlet	20	29	28	22	269*
Farm ponds	–	82	39	5	25
Dugout ponds + sunken ponds	29	35	15	21	2 + 2
Percolation tank	1 (repair)	1	1	1	1
Check wall	–	–	2		–
Check dam	3	1	16	2	1
Repairs to check dams	2	6	–	3	–
Horticulture (acres)	47	209	159	65	6200+
Fodder development	–	–	43	8	–

Note: *Figures are in cubic metres. + Figures in number of plants
Source: Respective DPRs and Project Completion report of Buttuvanipalle

rains during 2017, received almost a year's rainfall in a matter of few days. The benefits from the interventions are evident across the sample villages. In most of the sample villages, majority of sample households reported benefiting from farm ponds (including dugout ponds) and dryland horticulture (Table 3.5). In Battuvanipalle majority of the farmers reported benefiting from RFDs followed by trenches and farm ponds. Across the size classes greater proportion of large farmers are benefiting from various interventions followed by medium and SMFs (Table 3.6). This may be due to their better access to other capitals and resources, which is reflected in the case of farm ponds and check dams. It is now planned to cover all the farmers under farm ponds, i.e. each and every farmer will have a farm pond or covered under a farm pond user group.

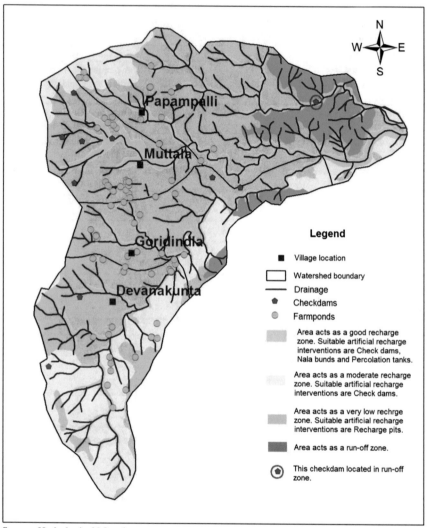

Source: Hydrological Mapping

Fig. 3.1 Available Artificial Recharge Intervention in the IWMP – Muttala Watershed
Source: hydrological mapping

3.5 Sustaining the Watershed Interventions and Beyond: Role of Institutions

Institutional evolution and strengthening are one of the main strengths of AF-EC. Its institutional approach is grounded in the philosophy that the role of local institutions is multidimensional and integral to overall village development. Over the years AF-EC has been promoting decentralized participatory institutions and given

Table 3.4 Status of watershed structures in the watershed villages (%)

Watershed	Working	Partially functional
D. K. Thanda (US)	78	22
Goridindla (MS)	80	20
Muttala (MS)	90	10
Papampalli (DS)	85	15
Battuvanipalli (NABARD)	100	0

Source: (1) Field visits and PRA/FGD methods; and (2) Watershed records/reports

Table 3.5 Sample households benefiting from different watershed activities village (watershed) wise (%)

Activities	D. K. Thanda (US)	Goridindla (MS)	Muttala (US)	Papampalli (MS)	Battuvanipalli (US)
Check dams	7 (18)	3 (8)	9 (23)	7 (18)	1 (3)
Trenches	9 (23)	7 (18)	16 (40)	7 (18)	11 (28)
Farm ponds	12 (30)	9 (23)	11 (28)	7 (18)	9 (23)
Dugout ponds	1 (3)	4 (10)	0	0	0
Mango plantations	4 (10)	11 (28)	6 (15)	0	6 (15)
Rock-fill dam	1 (3)	0	1 (3)	0	26 (65)
Stone bunding	2 (5)	6 (15)	4 (10)	7 (18)	1 (3)
Check walls	0	0	2 (5)	0	0

Source: field survey

the responsibilities to them. Involvement of village elders and PRIs from the beginning of the process has been the unique feature of the broader community involvement and participation. From the beginning AF-EC has been adopting a demand-driven approach to developmental interventions, i.e. communities are expected to demonstrate their collective and cooperative strengths in order to qualify for watershed interventions. AF-EC has a strong culture of promoting and facilitating CBOs that are committed to the practice of mutual cooperation, ownership of the project, and collective problem-solving and helps local leadership emerge.

AF-EC approach was not limited to creating programme-specific institutional arrangements that cease to function beyond the programme. The institutions need to continue beyond the programme to sustain the infrastructure and new activities and interventions over the years. In the case of WSD, this approach could be termed as watershed ++, as watershed+ is limited to livelihood enhancement. Integration of landless and other households depending on nonfarm activities into the institutional set up (like WDC or common interest group (CIG)) broadens the scope and responsibility of the institutions. This broader approach is reflected very much in the various interventions initiated in the watershed villages that go beyond watershed guidelines. For instance, resilience-building interventions are being introduced in order to make farming viable, while nonfarm activities like garment making, petty

Table 3.6 Sample households benefiting from different watershed activities watershed/village wise by economic category (%)

WS activity	S&M					Medium					Large				
	1	2	3	4	5	1	2	3	4	5	1	2	3	4	5
CDs	2 (11)	1 (5)	1 (6)	2 (13)	0	3 (18)	2 (13)	4 (27)	2 (13)	0	2 (40)	0	4 (44)	3 (30)	1 (7)
TRNC	2 (11)	4 (20)	4 (25)	2 (13)	3 (16)	6 (35)	2 (13)	7 (47)	2 (13)	2 (29)	1 (20)	1 (25)	5 (56)	3 (30)	6 (43)
FPs	6 (33)	6 (30)	3 (19)	0	0	4 (24)	1 (6)	3 (20)	3 (20)	3 (43)	2 (40)	2 (50)	5 (56)	4 (40)	6 (43)
DoPs	1 (6)	2 (10)	0	0	0	0	0	0	0	0	0	2 (50)	0	0	0
HRT	0	7 (35)	2 (13)	0	3 (16)	3 (18)	2 (13)	1 (7)	0	1 (14)	1 (20)	2 (50)	3 (33)	0	2 (14)
RFD	1 (6)	0	0	0	16 (84)	0	0	1 (7)	0	5 (71)	0	0	0	0	5 (36)
SB	1 (6)	4 (20)	2 (13)	2 (13)	0	1 (6)	1 (6)	2 (13)	5 (33)	1 (14)	0	1 (25)	0	0	0
CW	0	0	1 (6)	0	0	0	0	1 (7)	0	0	0	0	0	0	0

Note: CDs = Check dams; TRNC = Trenches; FPs = Farm Ponds; DoPs = Dugout Ponds; HRT = Horticulture (mango); RFD = Rock-fill Dams; SB = Stone Bunding; CW = Check Walls.; S&M = Small & Marginal Farmers; 1 = D. K. Thanda; 2 = Goridindla; 3 = Muttala; 4 = Papampalli; 5 = Battuvanipalle
Source: Field Survey

business, driving, etc. are being promoted and supported. And the livelihood portfolio keeps changing as per market requirements and local community demands. Women are the main focus in all these activities. This is grounded in the AF-EC philosophy that enhancing financial strength of women results in effective and sustainable household welfare.

In all the sample villages, the institutions, viz. WDCs and UGs/CIGs, continue to function even after completion of the watershed project. Another unique feature is the creation of WDF through community contribution. While this was a non-starter in number of cases across the state, AF-EC ensured that contributions are made genuinely and managed by the WDC. In fact, these bank accounts are still active with substantial balances. Further, these institutions are linked to the new initiatives like *Sasya Mitra* groups (SMGs) and MACS.

Sasya Mitra literally translates as friends of plants (and trees). SMGs are farmers' groups that plan, implement, and monitor the project. SMGs function as WDCs in non-watershed villages. AF-EC has been promoting SMGs in all their programme/project villages. There are 856 SMGs functioning in 214 project villages, and in 16 watershed villages, WDCs are functioning. Each SMG consists of 25 members representing the families, of which 13 are women members of the families. Members of each SMG elect a convener and a co-convener. Convener is a woman by policy. Each village has four SMGs, viz. two SMGs representing rain-fed farmers, one representing irrigated farmers, and one representing landless labour households. SMGs are federated at the village level called *Gramasasya Mitra Samakhya* (GSMS). The GSMS are federated at *Mandal* level called *Mandal sasyamitra samakhya* (MSMS). And the MSMS are federated at the project level called Apex *Sasya Mitra Samakhya* (ASMS) (see appendix Fig. 3.1A for the structure).

AF-EC, in collaboration with all the groups from village level to project level, prepares village wise yearly action plans with budget allocations and circulates to all the SMGs, GSMS, MSMS, and ASMS. Each SMG meets once in a month to discuss the implementation of the village annual plans and programme activities, savings, internal lending of saved money, and accessing government programme. SMGs select deserving and eligible beneficiaries for incentive-based activities considering their eligibility, interest, and ability to implement the activity. Gender and social equity is ensured during the selection by giving high priority to women, SC/ST/BC communities.

AF-EC has also been promoting a habit of thrift and mutual cooperation among the SMG members. A practice of mutual cooperation was promoted to reduce the cash transactions in agriculture activities and improve fraternity among the members. About 350 SMGs started small savings of Rs.50 to Rs.100 which was being used for lending small amounts of money to the needy members of the SMG. These small loans helped the members in meeting small but urgent expenses like the treatment of illnesses, paying children's education fees, purchase of grocery under Public Distribution System (PDS), etc. So far, an amount of about Rs. 3.5 million is saved and revolved benefitting about 7500 families every month. This has strengthened women's participation in decision-making at SMG level as well as household level.

GSMS is the village-level federation represented by conveners and co-conveners of each SMG in the village. GSMS is the focal point in the village and has been actively involved in planning, implementation, and monitoring of programme activities at village level. Each GSMS meets once a month. The major responsibilities of GSMS include allocation of incentive-based activities to SMGs, monitoring the output and use of output, and payment of incentives as per the guidelines. The management of the commonly owned equipment like sprayers, sprinklers, etc. provided by AF-EC is also the responsibility of the GSMS in each village. STOs facilitate these meetings.

There are eight MSMSs functioning in eight Mandals in the project area, represented by the members of GSMS. The MSMS meetings are facilitated at Mandal level by area team leaders (ATLs) and agriculture extension officers (AEOs). MSMS provides authentic feedback on the relevance and effectiveness of various activities and also suggests improvements in activities and their implementation. MSMS members also participate in participatory monitoring process conducted once during each cropping season. The teams of MSMS members visit randomly selected villages and monitor the progress and impact of project activities. They are also involved in planning, implementation, and monitoring of activities at the *Mandal* level. The MSMS played a key role in drawing the attention of government officials at Mandal level on the issues like timely supply of subsidy seeds distributed by the government during contingency cropping periods. They negotiate and lobby with the officials of the department of agriculture at the *Mandal* level and ensure timely supply of seed to the farmers.[2] The MSMS members also play an important role in organizing *Mandal* level awareness campaigns such as Drought and Desertification Day, World Water Day, International Women's Day, etc.

ASMS is represented by members one each from eight MSMSs, one each from five WDCs, and five progressive farmers who have a passion for sustainable agriculture. ASMS meets once in 3 months. It provides inputs in the planning process and provides feedback on the implementation of various programme activities and on the outcome and impact of programme activities. Further, it also discusses policy gaps and makes representations to the government authorities on policy issues. ASMS played a vital role in bringing to the notice of government the usefulness of cement lined farm ponds in protecting the tree crops as well as annual crops. ASMS is emerging as a strong forum for representing the issues of dryland farming and farmers to the government officials at the district level.

In the case of NABARD watershed, MACS was promoted by the VWDC. MACS took over the responsibilities of VWDC to ensure post-programme sustainability along with other activities following the legal and administrative procedures as per the MACS act. This arrangement ensured peoples ownership and responsibility for effective post-project sustainable management. MACS manages the WDF and uses it towards proper and timely maintenance of the watershed assets created under the programme. Besides, MACS manages the livelihoods fund released by NABARD, which is utilized as revolving fund (RF) on sustainable basis for enhancing livelihoods of the poor. MACS has been formed in NABARD watershed villages of Battuvanipalli, Mallepalli, and Garudapuram, for the purpose of managing credit and thrift, promoting alternative livelihoods and maintaining structures for improvement of natural resources. In the last 5 years, 1748 households have accessed credit from RF to the extent of Rs.224.62 lakhs for alternative livelihoods in the three MACS.

Battuvanipalli MACS has emerged as a model cooperative in AP, with 134 households taking loan for different purposes/activities. Out of which 12 (9%) members have taken for agriculture, viz. for purchasing of inputs, borewells, motors, and drip materials. As many as 114 (85%) members have taken for IGAs, viz. purchasing of *milch* cattle (buffalo, cows), sheep, goat, air compressor, animal business, auto works, carpenter, *dhobi ghat*, dish materials, fertilizers business, hotel, mechanic shop, purchasing of paddy (business), painting works, petty shop/business, rice business, saree business, stone cutting, tractor works, vegetable business, wood works, etc. Four members (3%) have taken loan for purchasing of two-wheeler, house construction, and marriage purposes. And the remaining 3% have taken loan for children's education.

---

[2] AF-EC also provides seeds to farmers and ensures timely sowing, though they may not be able to meet the entire seed demand in some seasons.

Photo 3.6 Farmer field schools (FFS)

While the MACS activities are limited to NABARD watershed villages, SMGs have a much broader mandate of overall development from village to district level. It is more like decentralized planning and implementation of programme and projects that build from village to district levels. ASMGs also influence policy by articulating the village-level needs. While this is proving to be very effective due to the continuous support from AF-EC, a systematic approach needs to be in place for scaling up to all the villages. The success of SMGs in the AF-EC programme villages is due to the high levels of awareness among the farmers regarding crop systems, technologies, livelihood activities, etc. And it is built over a period along with number of programmes and interventions including WSD.

For instance, while farmer field schools (FFS) are used in various programmes for building capacities and awareness about specific interventions, AF-EC has made FFS integral to agriculture development in all their villages. FFS in these villages serve as extension services in a more practical manner (Photo 3.6). FFS are conducted by STOs with the objective of increasing farmers' awareness of life cycles of crops, pests/diseases, life cycles of friendly and enemy insects, non-chemical ways of controlling pests and diseases, etc., through "observing, discussing, learning, and doing" approach as a group. FFS also experiment and demonstrate drought mitigation technologies and practices like drought-tolerant crop selection suitable for the soil, bio-nutrient management, protective irrigation during dry spells, etc.

In each village one standing crop is selected for organising the FFS. A small piece of land is left in the selected farm as control plot in which farming was done as per the farmer's conventional knowledge, while the remaining plot is used for experimenting new methods, crops, technologies, etc. FFS group is formed with 15–20 active and interested men and women farmers. Other SMG members and enthusiastic villagers are also encouraged to participate, discuss the outcome, suggest measures for improvement, and ultimately improve their knowledge and skills on crop management, drought management, and sustainable agriculture practices. "Field Days" are conducted in select fields just before harvesting of crop where hundreds of farmers come together and observe, listen, and discuss various practices followed including a cost-benefit analysis of demo plot vs control plot in managing the crop.

A new institutional initiative was introduced on a pilot basis during 2014–2015 in order to further strengthen the equity. The pilot initiated 7 rain-fed farmers' cooperatives (RFCs) with 170 farmer members with an objective to explore how SMFs can achieve livelihood security in the drought prone Ananthapuramu district. The cooperative strategy includes (i) reducing cost of cultivation by encouraging mutual cooperation among the members, (ii) enhancing farm productivity by promoting intensive drought mitigation measures, (iii) promoting sustainable agriculture practices, and (iv) diversifying livelihood portfolio to generate additional income of the rain-fed farmers by integrating collective farming, off-farm, and nonfarm livelihood activities on a collective basis.

Savings and credit are the basic principle even in RFCs. With monthly savings the 7 pilot RFCs have built a capital of Rs. 3,74,250 for meeting small financial needs of farm activities. Of the seven RFCs, four cooperatives have initiated off-farm and nonfarm livelihood activities like rearing lambs, harvesting and semi-processing of tamarind, and starting a seed business. Three RFCs took uprearing of ram lambs and were successful in running the collective business. Each of these RFCs made a net profit ranging from Rs. 26,000 to Rs. 74,000. The members felt that the amount of net profit was not at the expected level due to severe shortage of fodder during the summer, leading to poor weight gain of the animals and the resulting lower price for the animals in the market. But, the members were convinced that this activity can be highly profitable in future with good rains and success of contingency crops that assures good fodder availability. And some of the RFCs that ventured into business activities such as tamarind and millet seeds have either made losses or marginal profits. RFC members are also provided with hands-on experience, both on internal group management and on understanding market dynamics in 2-day training programme. The topics included principles of collective business activities and the opportunities and challenges associated with farmers' collectives. The experiences of some of the cooperative members who took up the collective business activities during the previous year were also used in the training.

3.6 Improving Drought Resilience: Cropping Systems and Technologies

AF-EC has been intensely involved in promoting cropping systems that can address drought mitigation. The choice of crops, how they are intercropped, and what goes into managing their cultivation have been key areas that AF-EC has been focussing on. AF-EC believes that much of the problems of drought can be minimized and mitigated through the right cropping system. A number of activities have been taken and promoted in this direction. These can be grouped under crop systems and practices and technologies and techniques.

3.6.1 Crop Systems and Practices

AF has been developing, promoting, and popularizing drought-resilient cropping systems with millets, pulses, and castor as intercrops/mixed crops through demonstration plots. Keeping in view the experience over the years, AF-EC demonstrated at scale the benefits of short-duration contingency crops like sorghum, horse gram, foxtail millet, green gram, and field beans which would, at least, provide the farmers with fodder and also with some marketable grain yields if conditions are suitable. As a drought mitigation strategy, contingency crops are introduced to overcome delayed rains. Government also provided sorghum and horse gram seeds to the farmers at a subsidized price in response to their demand. This helped overcome the fodder crisis. AF-EC's consistent efforts on popularizing these regular and contingency crop models brought in a positive trend with a large number of farmers. Farmers are shifting from groundnut monocropping to multiple and mixed crop systems and contingency crops whenever rains are delayed. There has been a sharp decline in the area under groundnut which is replaced by millets, pulses, and castor. In fact, the area under these three crops has gone up from 5 to 20%, as farmers are shifting to multiple and mixed crops (Photo 3.7).

Farmers view that they are able to cope with drought due to the shift from mono-crop culture to mixed/multiple cropping systems. Earlier the entire region was predominantly cultivating groundnut with red gram as an intercrop in rain-fed conditions. Now majority of the rain-fed farmers are cultivating bajra, korra (foxtail millet), jowar, cowpea, red gram, castor, and groundnut as mixed or single crops. Mixed and multiple cropping is a proven drought mitigation strategy (see Box 3.1).

Rain-fed tree crops are being promoted as an integral part of drought-resilient farming system. It would assure some income even in a drought year as the tree crops are more drought tolerant. However, the greatest challenge is to water the

---

Box 3.1 Mixed Cropping: A Climate Adaptation Strategy

Mixed cropping is a traditional farming practice. Farmers used to take as many as nine crops (Navdanya) at a time, which include cereals, pulses, oil seeds, and other crops. These provide all subsistence needs of the family. As some crops fix the nitrogen in the soil, some draw different kinds of nutrients and soil moisture from different layers. It creates a good balance in soil nutrient and moisture management. As each crop matures at different time and have different moisture requirements at different stages. One or two crops may fail due to moisture stress or untimely rainfall. The remaining crops will survive and provide at least some returns to the farmers. All crops together fail very rarely, may be once in 40 or 50 years. Bringing back this mixed cropping approach, farmers became less vulnerable to droughts. Further, as crops stand for longer period in the field, the soil exposure to wind erosion is minimized. Mixed cropping is also highly labour intensive. It is one of the best strategies to meet climate related uncertainties.

Photo 3.7 Mixed and multiple crops

Photo 3.8 Rain-fed tree crops

plants for the first 3–4 years for initial establishment and a long gestation period of about 5 years before the farmers can see any income from trees. AF-EC had designed four tree crop models – multiple fruit tree crops (MFTC), integrated farming system (IFS), bio-intensive farming system in rain-fed agriculture (BIFSRA), and afforestation/tree crops on waste lands and common property areas (Photo 3.8). AF-EC has demonstrated these models in about 80 ha. belonging to 80 farmers under rain-fed conditions and lobbied for their scaling up under MGNREGS. The government consequently came up with a policy for promoting rain-fed tree crops under MGNREGS and took it up on a large scale in the district. Due to this initiative, about 1700 ha. of mango and other tree crops came up in the project area including in the watershed villages benefitting over 1500 farmers, particularly the SC/ST and other small and marginal farmers. In order to avoid duplication, AF withdrew from promoting new tree crops. AF shifted its focus from promotion to supporting the farmers with extension services protecting the young tree plants, pest and disease management, productivity enhancement and market intelligence, etc., through mass campaigns and on-the-field demonstrations of pruning, fertilizer management, pest management, and productivity enhancement techniques.

3.6.2   Technologies and Techniques

Managing moisture stress to save the crop is the main challenge in Ananthapuramu district. Apart from crop systems like contingency crops, AF-EC has been promoting technologies and techniques in order to address drought. The initiatives include (i) timely sowing of crops under insufficient soil moisture condition, (ii) providing protective irrigation during prolonged dry spells, and (iii) promoting farm implements for drought mitigation and drudgery reduction.

In the climate change scenario, monsoons are delayed and the sowing window is shrinking. Given the short span of 90–120 days of crop growing period, timely sowing becomes critical. Further, given the low moisture holding capacity of the soils, fast sowing before moisture evaporates is important. In view of this AF-EC has identified and improvised the following equipment and technologies for timely and faster sowing (Photo 3.9).

3.6.3   Anantha Planter

Originally a tractor-drawn groundnut planter developed by the Regional Agricultural Research Station (RARS), AF-EC has remodelled the planter to make it conducive to the sowing of millets and pulses (Photo 3.9). The implement has a seed-metering mechanism which maintains seed to seed distance, the seed sowing rate, and has a

Photo 3.9 Technologies for timely sowing and efficient water use

component to automatically close furrows after sowing. It is four times more efficient than traditional planters.

3.6.4   Aqua Seed Drill

Anantha planter has been further improvised to provide water also along with the seed by mounting 200 litre capacity water drums with aligned pipes to provide water along with the seed. It is designed to sow in time in the absence of timely rains in the sowing season, i.e. June/July. Once seed along with water is sown, the seed would germinate and survive for about 2 weeks to catch up with subsequent rains.

3.6.5   Planter + Tanker

The method comprises of a planter which is fitted to a tractor-drawn tanker, wherein water is released through pipes simultaneous to the sowing. This is particularly efficient for red gram and castor crops requiring wide space between rows like 5–6 feet and ensures healthy germination and growth.

3.6.6   Watering Furrows

The technique involves the ploughing of furrows 5–6 feet apart, for castor or red gram, and watering the furrows followed by manual dibbling of seeds and closing the furrows. A water tanker of 5000 litres capacity is required for sowing castor and red gram on 1 acre of land. Shallow plough furrows were opened every 6 feet, and seeds were manually dibbled in the furrows. Water drawn from outside in a tanker was let in the furrows using pipes, and the furrows were closed with a plank. The method was demonstrated on 46 ha. of land belonging to 100 rain-fed farmers in the project area. This method works well provided it rains within 20 days after sowing. AF-EC has been trying to work out possible improvements.

3.6.7   Protective Irrigation

Providing protective irrigation during prolonged dry spells is as important as timely sowing. Over the last few years, AF-EC has been developing and demonstrating practices and techniques to mitigate moisture stress in order to save the crops. Studies have shown that more than 50% of the droughts are due to one dry spell of

Photo 3.10 Protective irrigation techniques

20–30 days. Prolonged dry spells during crop season cause moisture stress and crops fail. About 80% of the crop failures (droughts) can be saved if two protective irrigations can be given during such dry spells, particularly at the crucial periods of plant growth. AF-EC is promoting number of techniques and approaches in this regard (Photo 3.10). These include:

(a) *Mobile Protective Irrigation Unit (MPIU)*

As a response to water scarcity, AF-EC has developed a working prototype of a lorry-drawn tanker with an attached mobile sprinkler and drip irrigation unit. The unit has a tanker capacity of 12,000 litres and is used regularly for field demonstrations of various protective irrigation methods. The units are feasible for a scale up considering the large number of tractors and trailers in Ananthapuramu. Using this method, AF-EC demonstrated impact of protective irrigation for red gram on 375 ha of land.

(b) *Lining of Farm Ponds*

Another practical and low-cost solution for ensuring protective irrigation is lining of farm ponds with cement and clay mixture to harvest and store rainwater on-farm for longer periods. While testing the concept in Ananthapuramu, AF-EC has found that it can retain water for 30–45 days as opposed to farm ponds which are not lined, which only retain water for 8–10 days due to high seepage and evaporation. This technique has resulted in a 20–60% increase in yields during 2015 season. AF-EC has been advocating farm ponds on a large scale, and demonstrations were conducted for 23,600 farmers during the last 4 years. Consequently, the government has taken up construction of farm ponds on a large scale under MGNREGS. It is the most inexpensive method, and the farmer has much more control over it. Enthused by its success, many farmers are doing it on their own now. This method can be easily replicated in other drought-prone regions also. AF-EC encouraged the farmers with mango and other orchards to access the government programme like MGNREGS to construct farm ponds on their fields and to line it with cement and clay mixture for storing and using rainwater for protective irrigation. The success of these efforts is clearly reflected in the farmers increasing demand for construction of such farm ponds.

While protective irrigation is a proven drought mitigation strategy (see Box 3.2), availability of water is the most constraining factor in the Ananthapuramu district.

Box 3.2 Success of Protective Irrigation

My neighbour gave protective irrigation only once to red gram and the crop survived. AF had supplied 100 pipes and an oil engine for facilitating protective irrigation and farmers immensely benefited from it. We were able to irrigate even far off fields with these pipes. Some people had given irrigation to red gram crop twice by transporting the water with the tanker. We want more pipes this year. AF-EC has provided pumps, sprinklers, and pipes. Some farmers also managed to mobilize pipes from government."

Paparayudu, Farmer, Village: Sanapa.

I sowed groundnut in 3 acres near the tank after initial rains in June. I invested Rs. 30,000 on this crop. There was some rain in August but after that there was no rain. I gave protective irrigation twice during the season using the tank water. I can recover my investment and earn some profit. My neighbour who did not give protective irrigation even once was unable to save the crop and lost his entire investment.

Nasa Obayya, SMG member, Village: Sirimajjanapalli

I sowed groundnut in 4.50 acres and invested about Rs. 40,000. My farm is slightly away from the tank and despite using pipes I could not get water. My neighbour has a bore well and he could share his water for one protective irrigation. Although my crop survived, I got very little yield, the pod is very small and the quality is not good. I hope to recover at least half of my investment when I sell the produce. Those who gave protective irrigation twice during the critical period were able to earn some profit or at least could recover part of their investment but those who didn't, had lost their investment.

Kanchappa, SMG member, Village: Sirimajjanapalli

AF-EC is exploring various mechanisms to overcome the water constraint. Apart from technical innovations like mobile protective irrigation units, lining of farm ponds and efficient micro-irrigation systems[3] (sprinkler and drip), AF-EC is promoting mutual cooperation among the villagers, where one farmer provides water

---

[3] With these systems, castor and redgram crops require only about 10,000 litres of water per acre per protective irrigation and groundnut requires about 40,000 litres per acre per protective irrigation.

from his borewell or farm pond, and in return the provider receives the water-related works in his field. This initiative not only helped the members in reducing the cash transactions in agriculture operations through exchange of labour, implements, and bullocks but also created fraternity among families in SMGs. This year, AF-EC has encouraged GSMS members to explore the possibilities of sharing of water by SMG members with irrigation facilities with neighbouring rain-fed farmers for providing protective irrigation during dry spells. Farmers with borewells positively responded to the idea. AF-EC is putting conscious efforts to share these experiences across the district to encourage more and more farmers to join hands in water sharing. Farmers can also purchase water from borewell owners at a cost ranging from Rs. 500 to Rs. 800, and many farmers are willing to pay for the water to save their crop. In Sanapa village about 50–70 acres were irrigated by sharing borewell water (see Chap. 5 on groundwater sharing). Here most farmers seemed to have given protective irrigation to red gram crop.

   AF-EC has been developing and demonstrating protective irrigation methods to all the participating farmers in the project. Across the project villages, 8912 rain-fed farmers saved their crops covering 9700 acres during 2016. Due to the success shown by AF-EC in saving the crops, the state of AP has made protective irrigation a state policy and allocated Rs. 160 crores in the budget to scale up protective irrigation measures in 2016–2017 covering the entire state benefiting 10 million acres of drought-prone land. Government will provide infrastructure, equipment, and support for 2.5 million farmers in AP.

3.7   Other Interventions

AF-EC has been constantly searching and researching on farm implements that could help small and marginal farmers to retain moisture while reducing drudgery, particularly for women. The most recent interventions among them are subsoiler for better moisture retention and cycle weeder for performing weeding operations with ease (Photo 3.11). Subsoiler is a tractor-drawn implement which is used to make a very deep cut (up to 80 cm depth) into the hard soil to enhance the depth of topsoil.

Photo 3.11 Subsoiler and cycle weeder

AF-EC tested it on 40 ha. with deep ploughing before sowing. It is expected that deep ploughing will help in absorbing more rainwater and retaining moisture for longer time than the normal and will improve deeper penetration of roots. This experiment is being continued before concluding its efficacy. This method goes against the sustainable agricultural practices like "zero ploughing" promoted by other agencies. It would be interesting to see the results.

Cycle weeder is a simple implement innovated by a progressive farmer in Karnataka state by using a bicycle wheel for smooth and easy movement of the weeder on harder soils for weeding and inter-cultivation. AF-EC took some MSMS members on exposure visit to Karnataka. The women MSMS members were quite impressed with the weeder. AF-EC adapted the design to suit local crops and got 70 cycle weeders manufactured by local skilled agri-implement makers. These weeders received excellent response from, particularly, women, small farmers, and wage labourers. Usually weeding operations, a heavy back breaking job, are done by women. Cycle weeders do not need constant bending and make the weeding less strenuous. A lot of farmers started to purchase these on their own once they have seen and experienced the benefits. They are also inexpensive costing only about Rs. 1500 each.

Thus, effective implementation is a combination of capacities, institutions, upgradation of technologies, and techniques and adapting to changing conditions. Capacities of the implementing or facilitating agencies and the communities are critical as well. Only able PIAs can capacitate communities appropriately. Institutional evolution needs to be visualized over long term rather than conceiving programmatic institutions. This is the main lacuna with the mainstream watershed institutions which cease to function after the project completion. Though institutions are created specifically for the programme, they need to continue and evolve as per the local needs. For instance, convergence of WDC into SMGs or MACs. Sustaining agriculture is central to the wellbeing of the communities and efficacy of the institutions. Adoption of new technologies and techniques plays an important role, especially in the context of climate change with increasing incidence of drought. Making communities drought resilient goes long way in this regard.

Given the harsh environmental and natural conditions, the contribution of agriculture to the household income is limited despite the fact that majority of the population depends on agriculture. Dependence on agriculture is mainly due to the absence of viable livelihood opportunities rather than by choice. Agriculture is increasingly becoming unviable in general and more severely in these regions. While strengthening the natural resource base, especially water and soils, is necessary to improve the viability and share of agriculture income, it may not be sufficient to ameliorate poverty in the given natural constraints. A multipronged approach is required to bring these regions out of the dire conditions. Strengthening the natural resources base through watershed interventions is a medium- to long-term option, enhancing the resilience of the communities is a short- to medium-term option, and increasing the share of agricultural income or reducing the dependence on agriculture is long-term solutions.

The AF-EC technical initiatives fostered with appropriate institutional support emphasizes the importance of both short- and long-term options. Given the fragility of the region in terms of natural resources and the associated financial fragility of the household economy, AF-EC has realized early on that natural resource-based interventions like WSD are a necessary but not a sufficient condition for improving the economic conditions of the region (Reddy et al. 2004). The watershed institutions along with other institutions not only sustain the programme interventions but also support the communities in various other activities that strengthen and sustain their livelihoods. Its present approach is comprehensive and integrated, viz. WSD (resource base) + savings and credit based and livelihoods support institutions (enhancing financial capital) + technical and policy support (drought-resilient extension support and new programme interventions). The combined approach appears to be more effective in realizing the programme impacts and livelihood benefits. This approach goes much beyond the now widely adopted watershed plus approach and could be termed as "watershed plus + plus" (WS^{++}). A comprehensive assessment of the actual impact of these interventions in selected watersheds is taken up in the next chapter.

Appendix

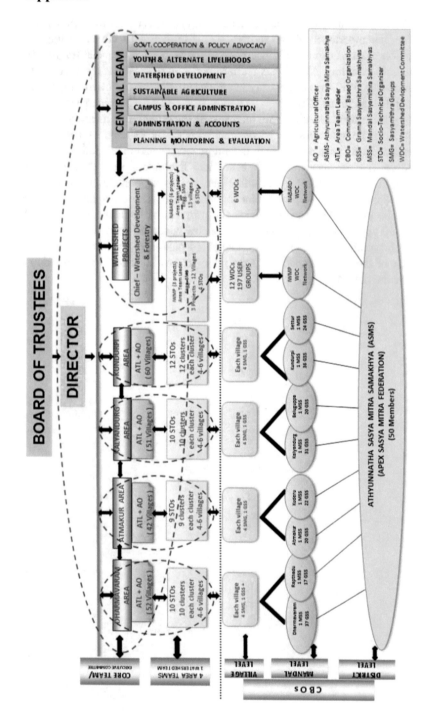

Fig. 3.1A Institution Structure of *Sasya Mitra Samakyas* (SMS)

Chapter 4
Sustainable Livelihoods and Resilience: Watershed Development and Beyond

4.1 Introduction

Impact assessment becomes complex when programme interventions are comprehensive and integrated. Attributing the impacts to a particular programme is rather difficult when multiple programmes/interventions by multiple agencies are taking place. This is a typical case with most of the villages in the region and more so in the case AF-EC implemented watersheds, as they adopt a multi-pronged approach for rural development. Their efforts are always to involve multiple agencies and their resources in an integrated fashion. This makes attribution very difficult, and hence the impacts assessed here are due to the overall interventions though watershed development (WSD) is central to all the complementary interventions. The programme in its entirety is termed as resilience and livelihoods strengthening interventions. While some of the impacts on natural resources could be specifically due to watershed interventions, others could be due to the package of practices, including WSD.

The impact assessment is based on the household- and village-level information collected on various environmental, socio-economic, and livelihood aspects. Besides, information is also drawn from the earlier evaluation reports where aggregate information was gathered. The assessment is focused on livelihoods and resilience. The sustainable rural livelihood framework of five capitals has been used to assess the livelihood impacts, and the resilience impacts are assessed using the households' abilities to cope with and adapt to drought. Since all the interventions are centred around and linked to watershed interventions, pre- and post-watershed situations are used to measure the changes. Besides, with and without watershed, situations are also compared. The assessment is carried out for each micro-watershed separately. Assessment is also done across socio-economic categories of households in order to capture the impact on equity.

The developmental activities as per the guidelines and community requirements were planned and implemented by the community with the PIA (AF-EC) taking the

© Springer Nature Switzerland AG 2020
V. R. Reddy et al., *Climate-Drought Resilience in Extreme Environments*,
https://doi.org/10.1007/978-3-030-45889-8_4

role of a facilitator. During implementation, convergence was ensured wherever required. Awareness building activities, regarding environmental degradation and the need for its prevention, restoration, augmentation of water resources, improving vegetative cover, along with alternative livelihoods through acquiring better skills and knowledge were carried out. A number of physical interventions to conserve soil and water and improve moisture were undertaken. The awareness building along with physical interventions had helped change/improve mind-set of the people to become active partners in the development interventions. Implementation in the selected watersheds was started in 2009–2010 and completed by 2015–2016, and the completion reports have been submitted. However, some additional works are being carried out, especially in the NABARD watershed.

4.2 Context

Ananthapuramu district is situated in a rain-shadow area where it gets very little rain and even when it rains it does not rain timely and distributed evenly to match the moisture needs of rain-fed crops. With land holdings of barely 1–5 acres, 90% of the farmers are small and marginal with limited access to irrigation. About 80% of them do not have borewells or other irrigation sources. During 2016, the district received the lowest rainfall, i.e., only 206 mm during crop season. More importantly the district experienced three consecutive drought years prior to 2017–2018 (reference year). Though this does not affect the before and after watershed comparisons (as before watershed was not a severe drought year), it would impact some of the indicators like livestock. For, households often sell out livestock during consecutive droughts, and they may take some time to acquire them back. This needs to be kept in view while assessing such impacts.

Under these conditions, the effectiveness of watershed interventions is limited though WSD is a necessary condition and serves as a natural resource stabiliser for this drought-prone region. Given the experience and core philosophy of AF-EC, the objectives of watershed implementation are similar and have to be implemented in a participatory approach involving all sections of people, though the watersheds were funded by different sources. Impact assessment is carried out on the basis of the objectives. The broad objectives include (1) conserve soil and water and improve vegetation (NRM); (2) promote horticulture and agriculture development; (3) provide employment for labourers during the programme and enhance rural employment opportunities on sustainable basis; (4) recharge groundwater for agriculture and drinking water for humans and animals; (5) provide credit for poor families to start off-farm and non-farm IGAs; and (6) capacity building to promote and strengthen institution building. Activities that suit the local context were taken up to fulfil these objectives. Given the low rainfall, poor access to surface water resources, and arid climatic conditions, the central activity of watershed is to harvest and store as much water as possible from rainfall, following a ridge to valley approach covering forest lands, other common lands, and private lands. Thus, the focus was mainly on water conservation, harvesting, recharge, storage, and efficient use.

All the interventions were planned and implemented in a participatory manner. AF-EC officials conducted an open meeting or *Gram Sabha* (GS) with all villagers and explained about benefits of WSD and also how it will be implemented in a participatory manner. Carrying out PRA exercises like social mapping and resource mapping helped encourage participation and consensus among the communities. Watershed development committee (WDC) was selected on consensus basis with 13 members and a chairman. Women, SC/ST, and other weaker sections were given importance in the selection of committee members. Few persons were selected to supervise the ongoing watershed works and report back to the committee. In few villages no chairman was selected due to political conflicts. Entry point activity (EPA) is the first intervention taken up under the WSD.

As many villages in the project area have excessive amount of fluoride in water resulting in health hazards, water purifying plants were taken up in all the watershed villages. Adinarayana, WDC member of Papampalli village of Muttala IWMP watershed, explains their importance: "Before the water purifying plant was installed, villagers had suffered with joint pains due to excess fluoride in drinking water (1600 ppm) and spent lot of money on the treatment. WDC installed reverse osmosis (RO) water purifying plant as part of watershed activities. We employed a local person to operate the plant on part time and pay Rs. 2,000 a month as salary. Villagers can buy 20 litre water can for as little as Rs. 2.00. The money is deposited in the watershed bank account, which is jointly operated by WDC members. The money is used for paying the salary of the operator and for maintenance. Now villagers are able to get safe drinking water at a minimum cost of Rs. 60 per month, which is far less than they were spending on buying water from the market or on treatment cost if they go to hospitals."

Actual watershed interventions could be grouped under four major interventions. They include SMC, RWH, crop systems and technologies, and savings and credit activities. SMC activities are on-farm activities like contour bunding, terracing, pebble bunding, etc., which help reduce runoff and erosion and helps improve moisture content. RWH interventions are usually on-stream interventions like check dams that impound rain water. Of late, RWH is also taken up at the farm level such as farm ponds, dugout ponds, and mini percolation tanks (MPT). Farm ponds have become a very important component in watershed activities. Rainwater is collected in farm ponds, stored, and being used to irrigate the rain-fed crops during long dry spells between rains. Farm ponds and dug out ponds can be constructed even in small holdings of 1 or 2 acres of land benefiting small farmers. When farm ponds are lined with cement, the water is stored for a longer period of time by preventing seepage. This technique has proved to be beneficial in saving the crops during crucial period of plant growth, especially in the case of groundnut – a major crop in this region. Check dams, percolation tanks, and other water bodies are mostly taken up on common lands or streams. They help recharge borewells.

Interventions pertaining to crop systems and technologies focused on shifting away from mono cropping and moving towards drought-resilient crop systems.

Horticultural crops along with multiple and mixed crop systems were promoted in the watershed villages. Tree plantation in farmers' lands, block plantation, and fodder development in community lands are other important components under this. AF-EC has selected tree species that are suitable for the climate of Ananthapuramu as well as beneficial to the farmers and local environment. This initiative has multiple benefits. Trees such as custard apple, mango, gooseberry, and tamarind (for fruits); glyricidia and pongamia (for biomass); neem (for organic pesticides); and ficus varieties like *ravi* (for fodder) are the usual species that are drought tolerant and provides benefits to farmers besides environment. Once the saplings grow, they provide assured income and other benefits to the farmer for a longer period of time. They provide green cover and fodder for animals and increase biomass. Trees such as mango, *jamun/black plum*, neem, and tamarind have been planted in an area of 5775 acres, benefiting 1520 farmers. WDC employs local villagers to water the block plantations. In Battuvanipalli village, they planted mango trees in 600 acres. At the same time, new technologies and practices were promoted to facilitate the shifts and drought management and enhance water use efficiency.

Savings and credit activities enable women to avail credit for enhancing or starting new IGAs. Shakuntala from Muttala village took a loan to enhance family business of incense stick making. "My two sons, daughter and I work at home making incense sticks. Although we have the equipment and skill, we were always short on cash for buying the raw material. I took a loan of Rs. 20,000 and it helped us in buying raw material. Now we are able to increase the production". Many women have taken loans for goat/sheep rearing, dairy and petty shops, etc. Nallamma from Papampalli village has taken a loan to purchase an auto and she operates it between Ananthapuramu town and Papampalli. *Ramalakshmi* in Goridindla village took a loan of Rs.30,000 to expand her sarees business. She says her position in the family has changed for the better after expanding the business. All of them feel dependency on moneylenders has greatly reduced due to the availability of credit. Besides the obvious financial gains, these activities have helped women become confident in their abilities and increase their self-esteem and their status within the family. Many farmers took loans to purchase farm implements. They rent them throughout the year to earn additional income. Similarly, MACS that are being promoted in the NABARD watersheds have multiple benefits (see Box 4.1).

4.3   Livelihoods

The combination of various interventions mentioned above is expected to have multidimensional impacts on households. In this section an attempt is made to assess these impacts systematically using the five capitals approach for a comprehensive understanding. These capitals include natural, physical, financial, human, and social. For, the nature of interventions reflects all these five capitals and hence likely to have an impact on them, and together they would influence the resilience of the households.

Box 4.1 Multiple Benefits from MACS

MACS is a cooperative, owned and managed by families themselves in the village to improve their economic and social wellbeing through credit, agricultural, and marketing activities. Farmers, artisans, and landless are members of MACS. They elect a governing body by consensus for operation and management activities of MACS. There is more involvement and participation from families as MACS is owned and managed by the villagers themselves. AF facilitated formation of MACS in Battuvanipalli in 2014. MACS committee with 15 members was formed with consensus of all villagers and 8 of them were women. Priority was given to SC/ST farmers as committee members. They formed by-laws for the operation of MACS, opened a bank account, and implemented many village development activities successfully. As of December 2016, MACS has a revolving fund of Rs. 24 lakhs with 152 loans and a fixed deposit of Rs.6.2 lakhs in the bank. Committee members meet every month.
Choudakka, *a woman committee member, explains, "Women take active part in discussion of loans and watershed works. We review loan applications and we consider applicant's credit history before sanctioning the loan. After sanctioning the loan, we physically go and inspect whether the applicant is using the funds for the income-generating activity she/he has put on the application. This ensures that they do not misuse the funds and have the capacity to repay the loan".*

Ramanjaneyulu, *Chairman of MACS eagerly shares, "We evaluate individual loan needs and we increased loan limit amount up to Rs. 50,000. We decided to do this as many villagers want seasonal agricultural loans or for purchase of goat/sheep. For this type of loan borrowers can pay only interest for 5 months and repay and entire loan in the 6th month. Short-term loans will facilitate loan repayments as they can make lump sum repayments when they sell produce or goat/sheep. During 5th to 9th of every month our MACS office looks like a bank where people stand in line to repay their loans".*

MACS formed various sub-committees for village development activities such as drinking water plant, temple development committee, school committee, etc. Village development committee (VDC) is providing school bags to children with money from village development fund. Working together for recovery of loans, resolving conflicts among themselves had a positive effect on strengthening and uniting the committee members.

Anand, *committee member, proudly shares: "A windmill company wanted to establish wind mills in farmers' lands for wind power generation and was willing to pay Rs. 9,000 a unit of wind mill. We negotiated with them and got Rs. 12,000 per unit for all farmers in our village without any mediator's help. Whereas farmers in other (non-MACS) villages could not get similar compensation". MACS of Battuvaripalli is thinking of processing and marketing agriculture produce.* Source:
Source: Personal Interviews

4.3.1   Natural Capital

The main objective of WSD is strengthening the natural resource base, which includes land, water, forests, and other commons. Given the priorities of the study region, some important indicators of natural capital are identified for assessing the impacts. These pertain to land and water. Watershed interventions could influence the quality aspects of land and land use. Land use is mostly in the form of cropping pattern changes and conversion of land from one form to another, i.e. waste lands to crop lands or seasonal crops to annual or perennial crops. Land use changes are linked to climatic conditions, markets, and access to water. Quantity and quality of water resources are influenced by watershed interventions. Both quantity and quality indicators are critical for drinking (human and livestock) and irrigation needs. Before and after the watershed, land values, area irrigated, and number of borewells clearly indicate positive change in all the villages, especially the watershed villages (Table 4.1).

All the sample households have reported increase in land values after the advent of WSD. Land values go up due to number of reasons, viz. secular market trends, increase in returns from farming, improved access to irrigation, and improvement in the quality of land. Watershed interventions could improve all but market trends. The increase in land values is substantially higher among the watershed villages (150–400%) when compared to control villages (107–124%) indicating the impact of watershed interventions. Battuvanipalli reported highest increase in land values (413%) despite a decline in area irrigated and number of borewells. The decline in area irrigated and number of borewells is limited to large farmers only. Large farmers reported decline in number of borewells in D. K. Thanda and Papampalle. The increase in the area irrigated is also the lowest in these two villages. This indicates that the impact of the interventions is favourable to small and marginal and medium farmers rather than large farmers. This is in the right direction as far as equity is concerned. Across the streams (within IWMP watersheds), midstream (MS)

Table 4.1 Changes in different indicators of sample households by economic groups (% change before/after)

Watershed	Small & marginal			Medium			Large			All		
	AI	LV	#BW	AI	LV	#BW	AI	LV	#BW	AI	LV	#BW
D. K. Thanda (US)	8	152	29	71	192	120	30	111	−43	21	152	26
Goridindla (MS)	22	284	120	17	198	−27	100	100	100	26	223	38
Muttala (MS)	23	178	50	37	157	117	190	177	225	77	164	119
Papampalli (DS)	4	77	50	14	155	0	6	195	−25	8	162	−8
Battuvanipalli	25	342	200	7	311	150	−3	486	−54	−12	413	−13
Kurlapalli (C-IWMP)	25	82	100	−2	137	−40	35	127	50	16	124	29
N. Puram (C-NABARD)	0	102	167	0	126	20	0	88	−67	0	107	14

Source: Field Survey

Note: *AI* area irrigated, *LV* land Value, #BW number of borewells

villages appear to be benefiting more with respect to all the three indicators, though downstream (DS) is expected to benefit more.

Soil quality impacts are measured in terms of erosion, runoff, and moisture retention. Most sample households reported moderate to good improvement in all the three indicators (Table 4.2). While very few sample households reported little (<19%) impact, majority reported good (>40%) improvement in most villages (except Papampalli). This is a clear impact of watershed interventions, as no such improvements are observed in the control villages. The improvements are relatively more in the upstream (US) and midstream (MS) villages, which is mainly due to the slope and quality of soils in these locations (ridge and middle). That is, US and MS locations have benefited more from the interventions. In Battuvanipalli, more than 60% of the sample households reported good improvement in all the three indicators. This could be the reason for the greater impact on land values in this watershed (Table 4.1).

Water quality is a major concern in these villages, especially for drinking purposes, due to excess fluoride in the groundwater. Communities used to face severe health problems like bone pains and disorders. Prior to watershed interventions, a number of villagers were buying water from outside (nearby towns). In all the watershed villages, mitigating the drinking water quality problem was taken up as first priority under the EPA. Under EPA RO plants were set up in all the watershed villages. This along with the improved water situation (reduced depth of groundwater table) the sources and access to drinking water have improved in all the villages (Appendix III Table 4.1A). A number of hand pumps increased in some villages. As a result, households spend less time in fetching water, i.e. come down by 50% in all the watershed villages.

As far as irrigation water is concerned, there are no serious water quality problems. People perceive that groundwater situation, especially borewells, has improved after the advent of watershed interventions. While there is no increase in number of open wells, about 28 defunct open wells were revived. A number of borewells have gone up by 26% across the watershed, and the depth of water table has improved by 12% (from 250 to 220 feet). Besides, other sources of irrigation and recharge

Table 4.2 Changes in soil quality indicators (%)

Watershed	Soil erosion			Run-off			Moisture retention		
	Little (<19%)	Moderate (<39%)	Good (>40%)	Little (<19)	Moderate (<39%)	Good (>40)	Little (<19)	Moderate (<39%)	Good (>40)
D. K. Thanda (US)	0.0	30.0	70.0	0.0	45.0	55.0	0.0	37.5	62.5
Goridindla (MS)	0.0	35.0	65.0	2.5	32.5	65.0	0.0	40.0	60.0
Muttala (MS)	0.0	27.5	72.5	0.0	22.5	77.5	5.0	47.5	47.5
Papampalli (DS)	5.0	60.0	35.0	5.0	47.5	47.5	0.0	57.5	42.5
Battuvanipalli	0.0	37.5	62.5	2.5	25.0	72.5	0.0	30.0	70.0

Source: Field Survey

Table 4.3 Aggregate changes in irrigation in the IWMP (Muttala) watershed

Character		Before	After	Difference	% change
Water resources	Open wells (nos)	66	65	−1	−1.5
	Borewells (nos)	250	315	65	26
	Depth to water table (feet)	250	220	30	12
Area irrigated (acres)	*Kharif*	891	1,216	325	36.5
	Rabi	460	629	169	36.7
	Summer	0	30	30	–

Source: Field Survey

structures like farm ponds, percolation tanks, check dams, etc. have also increased in the watershed villages. Due to the RWH structures along with SMC works, almost all the borewells are functioning during *Kharif* and *Rabi* seasons though some are drying up during summer season (see Box 4.2). These dried-up borewells are recharging during *Kharif* seasons. Farmers are moving towards less water-intensive crops (ID crops) and also adopting micro-irrigation systems. The net result is that area under irrigation has gone up by more than 36% in both *Kharif* and *Rabi* seasons (Table 4.3).

4.3.2   Land Use Changes

The mono-crop culture of groundnut is one of the major causes of severe distress conditions in Ananthapuramu district. AF-EC firmly believed that perennial tree crops grown under rain-fed conditions will mitigate drought and as well as influence the local ecosystem of the watershed area. Rain-fed trees like mango, custard apple, chikku, tamarind, jamoon, etc. that are known for their resilience in harsh drought conditions could provide secure income sources for the farmer. Additionally, these trees offer enough green cover (which is much wanted) as source of instant biomass and serve as fodder for small ruminants. A farmer's experience with tree crops: *"Our land used to get exposed to sun, wind and rain round the year reducing its fertility. Now because of the green cover and the biomass our land started recovering. Besides, mango we continue to get our regular seasonal crops as intercrops in the same field"*.

A combination of support measures from AF-EC under the WSDP has helped the farmers to overcome the constraints faced by the rain-fed horticulture farmers. These constraints include (a) long gestation period, (b) high initial investment, and (c) protecting the trees in the absence of assured source of water. The support assured under watershed programme for initial investment and particularly for pot watering encouraged the farmers to come forward to take up this initiative. Besides, farmers were introduced to mixed cropping with horticultural crop between the rows for 5 years or so, to avoid income losses. Regular extension support for proper crop management and yield enhancement in the form of individual interaction and

Box 4.2 Impact of Mini Percolation Tank in Battuvanipalli

An MPT was formed in the survey no. 306, to harvest the run-off of the surrounding lands. Farm bunding was also formed in the same survey number, to arrest the erosion and to enhance the retention of soil moisture, and to drain out excess run-off, stone outlets were formed. About 2000 plants were planted on the bunds. The survival rate was 70% (plantation register). Stone bunding was also constructed in the same land, where the slope of the land falls from the range 5–6% to minimize the erosion.

After the treatment of the above land (survey no. 306) and the formation of the MPT, the period of retention of soil moisture has been increased substantially and the crops too sustained during long dry spells after the treatments; the borewell, which is situated at the downstream side of the structure, got recharged and water level raised to 30 feet from 10 feet during the monsoon period.

A total of nearly 9 acres of land was brought to cultivation under the borewell. Paddy was cultivated in 1 acre and 5 acres of dryland horticulture and expected to generate Rs. 50,000 income. The farmer G.C. Ramanjineyulu started cultivating vegetables such as tomato, chilli, brinjal, etc.; he had earned nearly Rs. 90,000 from tomato crop and from chilli Rs. 15,000. An open well was also recharged at the downstream side structure in the same survey number, and farmer started cultivating tomato and earned Rs. 60,000. In the same survey number, in the fields of Anand Kumar, a borewell was get recharged by the MPT and supports cultivation of paddy (0.50 acres) and horticulture (5 acres with 250 plants). Together, MPT benefited more than 15 acres of land and supported high-value crops.

Source: Interviews with Beneficiaries

trainings provided by AF-EC has been very helpful for the farmers. And the convergence with government and other agencies like rural development trust (RDT) has helped expand the area under dryland horticulture, which provided the much-needed demonstration effect.

Land use changes in terms of cropping pattern changes is one of the significant and transformational impacts of the recent interventions. The changes observed in the region go against the conventional wisdom on watershed impacts. A number of studies have observed that watershed interventions result in a shift towards water-intensive crops like paddy due to improved water resource status. Despite the improved water situation in the sample watersheds, there is a clear shift away from paddy and groundnut. Prior to the interventions, farmers were used to mono-crop, viz. groundnut and paddy. Now farmers grow a number of crops including horticultural, millets, and inferior cereals. The average number of crops grown by a household has increased in all the sample villages (Table 4.4). It may be noted that large farmers grow a greater number of crops when compared to their counterparts. This may be due to their larger farm size. The shift is more or less uniform across the size

Table 4.4 Average number of crops grown by the sample households

Watershed	Small & marginal		Medium		Large		All	
	Before	After	Before	After	Before	After	Before	After
D. K. Thanda (US)	1.3	2.0	1.8	2.0	1.8	2.3	1.6	2.1
Goridindla (MS)	1.5	1.5	1.6	1.9	1.8	1.9	1.6	1.8
Muttala (MS)	1.0	1.3	1.5	1.7	1.0	3.0	1.4	2.0
Papampalli (DS)	1.6	1.6	1.8	1.9	1.2	1.8	1.5	1.8
Battuvanipalli	1.0	2.0	2.0	2.0	1.5	2.0	1.5	2.0
Kurlapalli (C-IWMP)	1.0	1.0	1.8	1.5	2.0	2.5	1.6	1.7
N. Puram (C-NABARD)	1.3	2.0	1.9	1.7	1.5	1.8	1.6	1.8

Source: Field Survey

classes in Battuvanipalli. Within the IWMP micro-watersheds, the increase in average number of crops grown is more among the DS and MS villages when compared to DS village. The increase is more in watershed villages when compared to control villages. Overall the farmers appear to have got convinced about the benefits of multiple crops. According to a farmer, *".... we don't want to depend totally on groundnut because it fails often....we are very happy that we raised two crops this year. It is wiser to spread the risk between two crops rather than resorting to gambling with one crop. We don't want to depend totally on groundnut because it fails often. Even our neighbouring farmers want to grow two crops from next year."*

AF-EC has been promoting dryland horticulture (mango, papaya, tamarind, etc.) in the region. Mango has become the most preferred crop in the watershed villages, which is mainly due to its ease in marketing. As a drought adaptation strategy, AF-EC is also promoting inter- and mixed cropping in a big way. As a result, area under dryland horticulture has increased from 90 to 780 acres and inter-/mixed cropping from 475 to 2578 acres. Dryland horticulture with perennial fruit tree crops like mango was undertaken under IWMP watershed scheme/component and necessary support extended to all the farmers on the criteria of soil suitability, interest towards horticulture, etc. In all the watershed villages, except Muttala, the soils are suitable for horticultural, especially mango crops. Requisite support in the form of plants, watering, and maintenance was extended by the PIA (AF-EC) initially for 3 years, which helped the survival of plants even in drought conditions.

Initially farmers were sowing mostly under rain-fed conditions. Farmers expressed that in the first phase, dryland horticulture was seriously affected by continuous drought for 3 years. Out of 5 years of WSDP implementation, 3 years were continuous drought years followed by normal rainfall year. Due to the watering and maintenance support during the first 3 years, the plants survived to a maximum extent (survival rate was as high as 90%). In the second phase, horticulture had normal rainfall conditions and all most all the plants survived. As the area under irrigation also increased due to watershed interventions, most of the irrigated farmers adopted drip method for irrigating their horticultural crops. Most of the farmers earn Rs. 6000 to Rs. 12,000 per acre per year on the mango crop. After the watershed interventions, 45 farmers have started growing mango crop on 188 acres as against 2 farmers growing on 5 acres prior to the watershed (Table 4.5). In the case of control villages, only in N. Puram (NABARD control), two farmers started growing mango on 13 acres. This clearly reflects the impact of AF-EC efforts in this regard, and the changes could be attributed to the watershed and the complementary interventions of AF-EC in these villages. In all the cases, groundnut/castor/tomato/red gram/jowar/green gram/horse gram is inter-cropped with mango. These crops provide the income as well as protect soils during the initial years, i.e. till mango yields start.

Most farmers with irrigation facilities are cultivating vegetable crops like tomato, green chillies, brinjal, etc. with drip systems. Before WSDP all the farmers were cultivating paddy, groundnut, and sunflower. Area under paddy has declined at the aggregated level (all watersheds) in both Kharif (−16%) and *Rabi* (−31%) seasons despite good rainfall and availability of water during the reference year, especially in the *Rabi* season (Table 4.6). Tomato is grown in all the villages, including control villages. Groundnut showed 7% increase during *Kharif* and 21% during *Rabi* season among the sample households, which could be attributed to good rainfall during the reference year. On the other hand, red gram and tomato have shown substantial

Table 4.5 Number of farmers growing and area under mango crop among the sample households

Watershed	No. of farmers		Area in acres	
	Before	After	Before	After
D. K. Thanda (US)	0	6	0	27 (4.5)
Goridindla (MS)	0	10	0	48 (4.8)
Muttala (MS)	0	9	0	42 (4.4)
Papampalli (DS)	0	5	0	28 (5.6)
Battuvanipalli	2	10	5	43 (4.3)
Kurlapalli (C-IWMP)	0	0	0	0 (0)
N. Puram (C-NABARD)	0	2	0	13 (6.5)

Note = figures in brackets are average area per farmer
Source: Field Survey

Table 4.6 Changes in area under crops in the watershed villages (all villages)

Crop	Kharif			Rabi		
	Before	After	% Change	Before	After	% Change
Paddy	2.5	2.1	−16	2.4	1.6	−31
Groundnut	2.7	2.8	7	2.3	2.8	21
Red gram	3.1	6.5	108	–	–	–
Tomato	1.9	2.6	35	1.8	2.2	22
Castor	5.0	2.3	−55	1.0	1.1	13
Green chillies	1.0	1.0	0	2.4	1.6	−31
Brinjal	1.0	0.8	−17	2.3	2.8	21
Green fodder	1.3	1.6	28	1.8	2.2	22

Source: Field Survey

Table 4.7 Changes in area under crops (major crops) among sample households across watersheds/villages

Watershed	Crop	Kharif			Rabi		
		Before	After	% change	Before	After	% change
DK Thanda	Paddy	1.8	2	13	1.9	1.6	−17
	Groundnut	1.8	2.4	36	0.5	2.3	350
	Tomato	1.5	1.6	5	1.3	1.4	12
Goridindla	Paddy	2.3	2.3	−4	–	–	–
	Groundnut	2.3	1.5	−36	2.2	2.5	14
	Tomato	1.7	3.4	102	1.5	2.3	56
Muttala	Paddy	2.9	4	38	2.0	3.0	50
	Groundnut	2.8	3.9	43	2.0	5.0	150
	Red gram	3.5	8.3	138	–	–	–
	Tomato	2	2.5	23	2.4	2.2	−8
Papampalli	Paddy	2.2	2.5	14	1.5	1.0	−33
	Groundnut	3	2	−34	2.0	2.5	25
	Tomato	2.1	1.9	−9	2.3	1.6	−31
	Green chillies	1	1	0	–	–	–
Battuvanipalle	Paddy	4.6	1	−78	3.3	0	−100
	Groundnut	3.3	3.5	4	3.3	2.1	−34
	Tomato	3.3	3	−11	1.7	2.8	66
Kurlapalli (C-IWMP)	Paddy	1.8	2	14	1.8	0	−100
	Groundnut	1	2.8	175	1.0	4.2	317
	Tomato	1.6	1.9	14	1.8	1.6	−9
N. Puram (C-NABARD)	Paddy	2.8	0	−100	–	–	–
	Groundnut	2.9	2.2	−24	2.7	2.0	−25
	Tomato	1	1	0	1.0	2.7	168

Source: Field Survey

improvement in area. The decline in area under paddy is the highest in Battuvanipalli (Table 4.7). The influence of watershed and related interventions of AF-EC becomes clear when IWMP micro-watershed villages and the control village are compared. The IWMP control village (Kurlapalli) has reported substantial increase in groundnut both in *Kharif* and *Rabi* seasons (Table 4.7). In the case of Battuvanipalli, there appears to be a demonstration effect in terms of crop changes, which has been reported. Besides, multiple seasonal crops like bajra, jowar, castor, etc. are grown as mixed crops with mango plantations in the watershed villages. And these changes are mostly confined to watershed villages. Extent of area under mango and mixed crops is to a lesser extent in the NABARD control villages, while they are absent in the case of IWMP control village.

Vegetable cultivation is short duration, low cost, and less water intensive. Though farmers apprehend about the price risk associated with vegetable crops, none of them have incurred losses so far. On the contrary, some of them get very attractive margins. Crop losses are minimum in the case of vegetables when compared to groundnut. A number of farmers have shifted from mono-crop to inter-crops and multiple crops methods/practices. This has also helped in reducing household consumption expenditure (food expenditure) and reducing the seed dependency and expenditure. In all the watersheds, farmers have moved from more water-intensive crops to irrigable dry (ID) crops, and most of them are using MI methods for irrigating and sustaining their crop during drought/dry-spell/moisture stress situations. Such shifts are not much in the case of control villages, which continue to adopt earlier crop patterns. Leaving lands fallow and migration during drought years is common in these villages. A more detailed account of cropping pattern changes across watersheds and control villages is provided in Appendix III (Table 4.2A, 4.3A, 4.4A, and 4.5A).

AF-EC has promoted inter-/mixed and multiple cropping along with the watershed implementation as part of their drought mitigation strategy. AF-EC has been demonstrating appropriate methods and practices suitable for rain-fed agricultural. They also provided extension support. AF-EC has designed and propagated ten models of inter- and mixed cropping systems with millets like foxtail millet, pearl millet, and finger millet, sorghum, and pulses like red gram, green gram, black gram, cowpea, and castor, which are drought resilient. These models are replacing mono-cropping of groundnut which involves high cost of cultivation. The cropping systems follow cost-saving and eco-friendly low external input sustainable agriculture (LEISA) principles and also improve food and nutritional security of the farmer families (Box 4.3).

Trainings and exposure visits to the successful/best practices areas of crop diversification were provided to the watershed committee (WSC) members and

---

Box 4.3 Benefits of Crop Diversification

Rameswar Reddy a farmer from Papampalli village used to cultivate paddy or maize in his 3 acres of land for a period of 8 years continuously. He never got sufficient income for the family, as he was investing more money on chemical fertilizers and pesticides. There was high incidence of pests and diseases due to mono cropping, resulting in high investment costs.

This year, motivated by AF staff, he went for crop diversification and has grown chilli, tomato, and some vegetables on his entire land (3 acres). He invested Rs.6000. He used the home-made bio-fertilizers and biopesticides on his vegetable crop. So, the investment on chemical fertilizers and pesticides, which he used to spend previously, became a saving/income for the family now. The net return on investment increased. No disease was found except for some leaf spot. He had a good crop of tomato and chillies, and till September end 2017, he has earned Rs.28,000 by selling chillies and tomatoes (in 4 months) and still can harvest tomato for another 2 months and chillies for 4 more months. This not only helped in fetching more than paddy/maze but also helped their family with nutritious food of fresh vegetables and a better quality of life.

Source: Personnel Interview

sub-committee members. According to a farmer, *"We are happy for the shift from mono-cropping to crop diversification to vegetable cultivation which gives higher return on investment than paddy. This in turn helped our family members to eat nutritious food with fresh vegetables. Thanks to AF-EC for suggesting crop diversification in time and for the supply of good quality seeds and the awareness building on LEISA and NPM practices"*. Another farmer expressed that *"my friends and relatives have praised me for taking a good decision by choosing to sow cowpea. My wife and daughters are happy to add one more food grain to our food stock. Above all it gave me good profits that helped me to recover"*.

The land use change as a strategy for drought mitigation under watershed programme in Battuvanipalli has motivated other surrounding villages to shift to new cropping pattern. This is reflected in the increased area under mango and vegetable crops in its control village, N. Puram. More importantly, the policy of the state government has been tuned to take up dryland horticulture as a programme for rain-fed famers under its MGNREGA programme. Now thousands of farmers across the district are accessing MGNREGS and converting part of their cultivated land to suitable tree crops. Thus, the impact of the interventions is not limited to target villages or watersheds; they are widespread reflecting the benefits from them. The ultimate success is in getting policy attention for scaling up at the district level. This may change the land use scenario of Ananthapuramu district, which is known for groundnut mono-cropping, with multiple crop systems in the near future.

4.3.3   Physical Capital

Physical capital is measured in terms of livestock holdings and assets. Assets include immovable assets like houses and cattle sheds and movable assets like agricultural implements, machinery, and household durables. Impact on livestock is assessed separately for total livestock and milch animals in value terms. It is expected that watershed interventions would increase the milch animals and small ruminants and reduce the draught cattle, as access to mechanization increases. Under watershed works and activities, dairy farmers were encouraged through increased production of green fodder at the household level. Area under green fodder increased by 20% in all the watershed villages due to the distribution of subsidized fodder seed to the irrigated farmers in order to promote stall feeding. Dairy farmers also benefited by way of fodder development in CPRs for open grazing during *Kharif* and *Rabi* seasons. Dairy was also encouraged directly, as dairy was one of the activities under IGAs supported by the livelihoods fund. In addition to this, AF-EC also provides livestock health and supported activities. According to a farmer, *"Thanks to AF-EC for arranging cattle health camps at the right time. I could save my cows, bullock and sheep from diseases because of this cattle health camp. It saved me a lot of expenditure"*.

However, the effectiveness of these interventions is not evident in the present scenario. This is mainly due to the three consecutive droughts in the region, which

Table 4.8 Changes in livestock holdings in sample villages

Watershed/village	Milch cattle		Draught cattle		Young stock		Small ruminants	
	Before	After	Before	After	Before	After	Before	After
D. K. Thanda	96 (19)	12 (9)	54 (24)	10 (5)	36 (12)	5 (4)	190 (6)	152 (7)
Goridindla	53 (15)	14 (10)	36 (17)	8 (4)	10 (4)	10 (8)	300 (4)	0 (0)
Muttala	126 (25)	35 (13)	38 (17)	6 (3)	54 (20)	18 (8)	560 (6)	83 (3)
Papampalli	15 (8)	50 (16)	16 (7)	4 (2)	5 (4)	30 (15)	298 (4)	200 (4)
Battuvanipalli	71 (25)	52 (22)	28 (14)	4 (2)	39 (25)	29 (20)	433 (9)	53 (4)
Kurlapalli (C-IWMP)	23 (13)	22 (13)	18 (8)	11 (6)	9 (7)	9 (8)	600 (11)	431 (9)
N. Puram (C-NABARD)	31 (12)	34 (15)	23 (11)	9 (4)	16 (11)	23 (13)	100 (2)	40 (1)

Note: figures in brackets are number of sample households with livestock
Source: Field Survey

adversely affected the household physical capital build-up in the sample villages. For, disposing livestock and movable assets is a known household drought adaptation strategy in this region. This is evident in all the sample villages where a number of households holding different categories of livestock and their numbers have declined substantially in all the sample villages (Table 4.8). Only Papampalli reported increase in milch cattle and young stock. The decline is more in the small ruminant's category in all the villages (details on cost of rearing and income from livestock are presented Appendix III Tables 4.6A and 4.7A). Increase in Papampalli is mainly due to the distribution of milch cattle by the RDT in recent years. This programme is now being extended to other villages. Besides, Papampalli has the highest area under fodder (8 acres) among the watershed villages, which supports the increase in milch cattle despite the three consecutive droughts.

Livestock population, measured in total livestock units, declined in all the watershed villages except Papampalli (Table 4.9). On the contrary, control villages have reported increase in livestock. In the case of milch cattle also, three of the watershed villages have reported decline. Papampalli reported substantial increase (335%) in milch cattle and Muttala has shown 19% increase. In most watershed villages, the decline in milch cattle is lower when compared to total livestock units. This is mainly due to the fact that households dispose small ruminants first followed by drought animals and milch cattle during drought years. Besides, increased use of machinery has reduced the dependence on draught cattle. This is reflected in the increase in the assets. Assets have gone up by more than 100% in all the watershed villages. The change in household assets is higher in the watershed villages when compared to control villages. Across the size classes, medium and small and marginal farmers have reported greater increase when compared to large farmers in most of the watershed villages.

Earlier evaluation reports also indicated an increase in milk production by more than 70% in the IWMP watershed villages. This is mainly attributed to the increased availability of green fodder and the decline in foot and mouth diseases due to AF-EC livestock health and supported services. There is also a shift in favour of hybrid

Table 4.9 Changes in different indicators (value terms) of sample households by economic groups (% change before/after)

Watershed	Landless			S&M			Medium			Large			All		
	TLS	MC	Asst	TLS	MC	Asst	TLS	MC	Asst	TLS	MC	Asst	TLS	MC	Asst
D. K. Thanda	100	100	46	−47	−68	158	−59	−82	132	−89	−92	85	−59	−79	122
Goridindla	0	0	61	−75	−33	193	−67	−60	151	−38	−100	239	−69	−54	165
Muttala	0	0	49	−81	−64	119	−71	−65	196	192	285	80	−25	19	123
Papampalli	100	100	94	255	100	155	40	124	93	230	496	85	137	335	100
Battuvanipalli	−54	−58	−9	−68	−48	229	90	177	241	11	59	233	−30	−4	188
Kurlapalli (C-IWMP)	−39	−56	165	53	57	123	1	16	45	67	153	76	13	8	93
N. Puram (C- NABARD)	100	100	65	−46	−11	157	105	148	125	43	−17	223	12	59	149

Source: Field Survey

Note: *TLS* total livestock, *MC* milch Cattle, *Asst.* assets

Vincreasing the milk yield. Dairy has now become a commercial activity, which supports household expenditure.

4.3.4 Financial Capital

The combination of natural capital and physical capital translates into financial capital at the household level. Financial capital is measured in terms of income from different sources like agriculture (crop income), livestock, employment, and other IGAs. Income from all these activities is assessed in terms of household income, savings, and debt of the households. It may be noted that getting accurate information on income and savings/debt is a challenge. We have tried to cross-check the information using different ways like income, consumption, activity-wise information, etc. However, comparative assessment would be more meaningful than looking at absolute figures. The average household income ranges between Rs. 84,000 and Rs. 1,41,000 across the sample villages (Table 4.10). All the watershed villages have a higher average income when compared to control villages, indicating the positive impact of watershed interventions. Of the watershed villages, Battuvanipalli (NABARD) has higher average income than IWMP micro-watersheds. Even the control village of NABARD has higher income, indicating that these differences could be due to the agroclimatic conditions rather than watershed implementation per se. Differences in average income could be noticed with in IWMP micro-watersheds. Average income is higher in the downstream (Papampalli) when compared to upstream and midstream villages. As expected large farmers have higher incomes when compared to medium, small, and marginal and landless households reflecting the importance of land in contributing to household income. Most of the village's household income matches with their general consumption (food, health, education, entertainment, etc.).

Agriculture is the single largest contributor to household income. Its contribution ranges between 40 and 68% across the villages (Table 4.11). Agriculture contributes more in watershed villages, especially in the IWMP micro-watersheds. In two of the

Table 4.10 Average household income across economic groups among the sample households (Rs./year)

Watershed	Landless	Small & marginal	Medium	Large	All
D. K. Thanda (DS)	63,500	88,333	90,532	147,000	90,140 (93,531)
Goridindla (MS)	75,083	90,400	163,687	270,000	109,020 (115,344)
Muttala (MS)	82,000	88,384	192,333	231,250	125,560 (136,908)
Papampalli (DS)	79,789	74,159	113,988	270,000	128,210 (157,156)
Battuvanipalli	76,717	100,958	140,834	215,385	140,740 (89,127)
Kurlapalli (C-IWMP)	73,625	42,905	122,077	151,000	83,840 (80,738)
N. Puram (C-NABARD)	56,444	78,361	147,167	262,500	118,460 (71,695)

Note: figures in brackets are household annual consumption
Source: Field Survey

Table 4.11 Changes income sources in the sample villages (%)

Source	D. K. Thanda		Goridindla		Muttala		Papampalli		Battuvanipalli		Kurlapalli (C-IWMP)		N. Puram (C-NABARD)	
	B	A	B	A	B	A	B	A	B	A	B	A	B	A
Agricultural (net)	47	53	57	55	39	68	68	52	63	55	42	40	44	55
Livestock	9	1	9	2	13	6	2	10	7	4	4	3	5	4
Farm Labour	19	21	18	14	9	5	9	10	12	11	24	23	21	15
Non-farm labour	8	12	4	5	7	3	4	7	4	4	10	14	5	5
NREGA	12	8	11	5	6	2	5	1	5	3	9	0	8	0
Employment$	2	3	1	12	2	7	6	12	6	18	7	14	7	14
Business	2	1	0	3	15	7	4	5	3	6	3	3	6	4
Hiring out#	2	0	1	0	9	3	0	4	1	0	2	3	1	0
Others*	0	0	0	4	0	0	0	0	0	0	0	0	2	4

Note: $ = include government and private; # = include bullocks, tractor and other implements, * = include village services, remittances, and CPRs. Source: Field Survey; B = before; A = after

four micro watersheds, contribution of agriculture has gone up after the watershed. Papampalli has recorded sharp decline in the contribution from agriculture (from 68 to 52%), which is mainly due to the increased contribution from livestock and employment. For, Papampalli is the only village that reported increase in income from livestock, while all other villages reported a decline (see Appendix III Table 4.8A and 4.9A).

On the contrary, the contribution of agriculture in the Muttala micro-watershed has gone up from 39% to 68% after the watershed interventions due to the decline in the contrition of livestock and other sources. Contribution of agriculture in Battuvanipalli declined due to the increase in the contribution from employment (govt. and private). Labour is the second largest contributor with 20–40% share across the villages. Interestingly, there is a universal decline in the contribution of NREGA to the household income. While this could be due to the convergence of watershed and NREGA works in the watershed villages, it is difficult to explain its decline to zero contribution in both the control villages. This may need specific probing. Across the farm size classes, contribution of agriculture is the highest for large farmers. Even for small and marginal farmers, agriculture is the single largest contributor in most villages (see Appendix III Tables 4.8A and 4.9A).

This emphasizes the need for strengthening the agriculture (natural resource base) in these regions. This is somewhat different from the situation in the land constrained (but water abundant) Indo-Gangetic plains of Uttara Pradesh (UP), Bihar and West Bengal, where the share of agriculture is lower (much below 50%) from that of labour and other sources (Reddy et al. 2017). The water constraining regions appear to have the potential for expanding and strengthening agriculture through appropriate water and moisture management. On the contrary the land constraining regions have been limited by a maximum crop intensity of 300%. In fact, some of the regions in the Indo-Gangetic plains have crossed 200%. That is, solutions to poverty in these regions need to be found out-side agriculture. Whereas, strengthening agriculture sector could still alleviate poverty in regions like Ananthapuramu (relatively more land but less water). Hence, public investments in agriculture continues to be an efficient resource allocation strategy for poverty alleviation in these regions.

The increased contribution of agriculture is reflected in the yield improvements of important crops over the period. Per acre productivity has gone up for all important crops, except castor (Table 4.12). Productivity increases ranged between 13 and 68% for groundnut, red gram and tomato (more detailed data on returns per acre and crop wise costs returns are presented in Appendix III Tables 4.10A, 4.11A, and 4.12A). These figures commensurate with the earlier impact evaluations of the same watersheds. This indicates the sustenance of the watershed impacts. Castor has recorded a decline in productivity in all the villages, except Papampalli and D. K. Thanda. This could be due to unfavourable climatic conditions and related pest incidence. Productivity gains are more among watershed villages when compared to control villages. Within the IWMP micro-watersheds, midstream villages of Goridindla and Muttala have performed better. Even the decline in castor productivity is more in the control villages when compared to watershed villages. Only in the case of groundnut, IWMP control village (Kurlapalli) has reported higher increase in productivity. Of late, groundnut is slowly loosing importance in the watershed villages. Productivity gains are more among small and marginal farmers when compared to large farmers.

As a result of increased productivity for most crops and also the returns from horticultural crops (see Box 4.4), average gross and net returns per acre have gone up substantially in all the villages (Table 4.13). The increase in net returns are more than gross returns indicating a decline in cost of production. In all the sample villages, except Goridindla, small and marginal farmers have gained more than large farmers. Within the IWMP micro-watersheds midstream villages, especially Muttala has gained more. Across the sample villages, while gross income per acres varied between Rs. 21,000 and Rs. 46,000 (Fig. 4.1), net income varied between Rs. 13,000 and Rs. 27,000 during the post -watershed period.

Apart from farm and livestock development activities, AF-EC has been supporting the communities with another off-farm or non-farm IGAs. These include

Table 4.12 Changes in crop productivity of important crops across sample villages (%)

Watershed	Groundnut	Red gram	Tomato	Castor
D. K. Thanda (US)	29	21	21	36
Goridindla (MS)	30	67	67	−1
Muttala (MS)	62	15	15	−13
Papampalli (DS)	17	33	33	25
Battuvanipalli	49	20	20	−14
Kurlapalli (C-IWMP)	68	13	13	−17
N. Puram (C-NABARD)	28	26	26	−100

Source: Field Survey

Box 4.4 Returns to Mango Crop in Battuvanipalli

AF-EC was trying to sensitise farmers on the importance of dryland horticultural crops in the wake of fast depleting ground water situation during the initial years of watershed interventions (2010–11). AF-EC's continued efforts influenced Mr. Thimmarayudu, who came forward to take up mango plantation on 6 acres of land. He convinced his father and brother and planted 450 mango plants on 6 acres. All the members of his family put in a lot of efforts in watering and protecting the plants from June 2011. They used their bullock cart to carry water in drums for watering the plants. They followed the suggestions of AF and RSO on mango crop management. Inspired by these efforts of Thimmarayudu, 75 farmers of the village had planted fruit plants covering 172 acres in 2 years from 2011 to 2013. And another 102 S&M farmers shifted to mango orchards on 92 acres in the year 2014.

Mr. Thimmarayudu earned about Rs. 1,36,000 from the first harvest of mango in 2016 season. He also gained from intercrops of groundnut, red gram, and horse gram during the 4 years. According to him, "I spent about Rs.50,000 on mango crop management and got a net income of Rs.86,000 with the first harvest. In the future the expenditure comes down to Rs.10,000–15,000 and the yield increases further which assures me of higher income". More and more tree crops are not only providing income to the farmers but also positively contributing to the environment.

Source: Personnel Interview with the beneficiary and data from AF-EC Office.

Table 4.13 Changes in gross and net farm income from crops (%/ acre)

Watershed	S&M		Medium		Large		All	
	Gross	Net	Gross	Net	Gross	Net	Gross	Net
D.K.Thanda(US)	160	237	133	217	12	17	112	162
Goridindla (MS)	199	255	63	86	280	454	101	137
Muttala (MS)	605	646	216	347	100	211	270	377
Papampalli (DS)	113	168	98	137	91	142	98	145
Battuvanipalli	301	387	177	195	112	97	162	169
Kurlapalli (C-IWMP)	205	532	39	54	−3	16	55	109
N. puramam (C-NABARD)	315	424	227	313	212	268	258	344

Source: Field Survey

diversified livelihoods for educated and undereducated rural youth by imparting requisite skills and competencies through its in-house driving schools and garment-making training centre and by establishing partnerships with vocational training institutes and potential employers from the private sector. The idea is to facilitate skilled employment or self-employment for undereducated rural youth, particularly those from landless families. It is also aimed at creating upward occupational

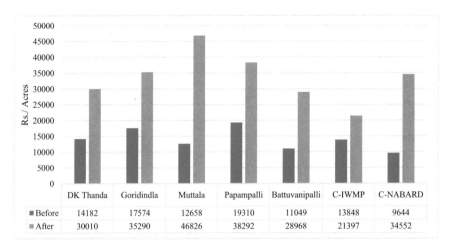

	DK Thanda	Goridindla	Muttala	Papampalli	Battuvanipalli	C-IWMP	C-NABARD
Before	14182	17574	12658	19310	11049	13848	9644
After	30010	35290	46826	38292	28968	21397	34552

Fig. 4.1 Changes in gross income per acre across the sample villages (Rs. /Acre)
Source: Field Survey

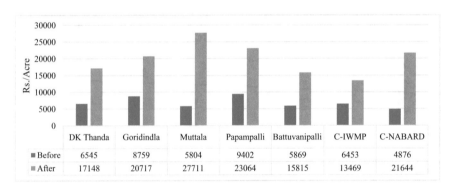

	DK Thanda	Goridindla	Muttala	Papampalli	Battuvanipalli	C-IWMP	C-NABARD
Before	6545	8759	5804	9402	5869	6453	4876
After	17148	20717	27711	23064	15815	13469	21644

Fig. 4.2 Net income per acre across sample villages (Rs./acre)
Source: Field Survey

mobility for the younger generation in contrast to traditional occupations like wage labour in agriculture. The focus is also on women, particularly in enabling self-employment towards economic freedom. Women were trained in tailoring and provided with advanced training on garment making. AF is exploring possibilities to facilitate linkages between the trained women and garment industry. Besides, driving school for rural youth and two-wheeler mechanic training are provided (Fig. 4.2).

The SHGs and the MACS have got boost under the WSD. These initiatives provide support to establish and run small IGAs (see Box 4.5). As a result, average savings as well as debt of the households have increased substantially over the period (see Appendix III Tables 4.13A, 4.14A, 4.15A, 4.16A, 4.17A, and 4.18A for details). The average debt has increased steeply when compared to savings in all the sample villages. The debt-savings ratios have gone up by 3–5 times in all the

Box 4.5 Better Access to Loans for Income-Generating Activities

Many women in these villages have availed the loan facility and benefited from diversified income-generating activities such as starting petty shops, selling vegetables, basket making, buying an auto, stonecutting, selling clothes, etc. Now they are more focused on income generation and asset creation rather than using the funds for family consumption. An important benefit is that they earn income throughout the year instead of depending on seasonal agricultural income.

Suvarna, who didn't have any supplemental income earlier, took a loan of Rs.30,000 and purchased two buffaloes. Now she sells 10 litres of milk every day to a local dairy and earns Rs.6,000 per month, out of which she pays Rs. 1700 as the monthly instalment towards loan repayment and also spends some amount on fodder. Rest she is saving for her children's education.

Manjula got the idea of selling sarees when she attended meetings outside her village and saw others doing it. "I borrowed Rs.30,000 from IWMP livelihoods funds and another Rs.30,000 from the bank. My husband and I went to Proddutur town and bought sarees at wholesale prices and I sold them here in the village. There is good profit in this business and I am planning to add bed sheets, etc. Some customers pay in instalments as they cannot afford to pay the entire amount in a lump sum". She brings out another interesting dimension of barter system when she says "Sometimes women work in our fields in exchange for part or full payment of a saree". She repays Rs.1800 and Rs.2000 in monthly instalments towards IWMP and bank loans, respectively.

Lakshmi Devi took a NABARD loan of Rs. 20,000 and a bank loan of Rs. 30,000 and purchased 15 sheep. Within 4 months she sold them for a profit of Rs.30,000. She plans to take another loan for the same activity once the current loan is paid off. This, she says, would not have been possible from agricultural income alone. Venkatehwarulu in Yerragunta took a loan and put some of his own money to buy a tractor and now he gives it on rent during the season, earning additional income.

Both IWMP and NABARD livelihood programmes lend loans at nominal interest rates of 0.25% (IWMP) and 1% (MACS) per month. The ease of borrowing at nominal interest rates and paying in monthly instalments has reduced the dependence on exploitative moneylenders. The dependence on moneylenders gives them a power to dominate the farmers economically, socially, and politically. Sujatha explains, "If we borrow from a moneylender we have to pay the entire amount at one go. It's difficult for us to accumulate the entire principal plus interest amount". Paying in monthly instalments has taught the women the discipline of savings and budgeting cash flow for various needs. Loan recoveries are about 75–80% in the project area.

Source: Personnel Interviews with Beneficiaries.

Table 4.14 Changes in debt-savings ratios of sample households across sample villages

Watershed/ village	Landless		Small & marginal		Medium		Large		All	
	Before	After	Before	After	Before	After	Before	After	Before	After
D. K. Thanda	0	1855	465	4520	486	6889	0	14,657	400	5912
Goridindla	405	2167	404	2250	771	3910	0	2619	350	2836
Muttala	0	3268	874	3399	1586	3034	0	4500	764	3469
Papampalli	197	2902	359	3256	2266	5393	0	3203	660	3702
Battuvanipalli	1732	4709	672	1132	2752	2507	65	956	460	1367
Kurlapalli (C-IWMP)	1750	4815	1199	4494	700	2342	1500	10,897	1252	3888
N. Puram (C-NABARD)	1887	2971	206	1528	158	2504	0	1398	255	1729

Source: Field Survey

villages (Table 4.14). In D. K. Thanda, the ratio has gone up by more than ten times. It may be noted that in all the IWMP micro-watersheds, large farmers have got into debt during post-watershed period. Similarly, landless households also got into debt in two micro-watersheds (D. K. Thanda and Muttala) after the advent of watershed. This could be due to two reasons, viz. (i) increased debt burden due to three consecutive drought years (demand) or (ii) increased access to credit consequent to WSD (supply). It appears that the combination of demand as well as supply might have pushed the debt-savings ratios substantially in the watershed villages. The debt-savings ratio is substantially lower in the NABARD watershed (Battuvanipalli) villages, where a mutually aided cooperative is successfully working.

4.3.5 Human Capital

Human capital pertains to the abilities of a household to use their human resources in different livelihood activities. The allocation of human resources across activities depends on the attributes of these resources. These attributes include age, gender, health, skills, and education. While some of these attributes like natural capital are given, some of the attributes like education and skills need to be developed over a period that requires efforts and investments. Human capital provides long-term livelihood gains and sustenance to the household like natural capital. Given that there are not many differences in terms of number of working people/dependents and gender ratios are not very different across the villages, here skills, education, and health are considered as indicators of human capital. The dependent – earner ratios range between 2:3 and 1:3 across the sample villages.

As discussed in Chap. 3, watershed villages get lot of support from AF-EC in enhancing the community's skills. These include not only watershed-related skills but also new technologies and techniques and alternative livelihoods. These skills are proving to be beneficial in sustaining the watershed interventions, improving farm productivity, and earning additional income. The support received from WSDP has strengthened the SHGs and mutually aided cooperative societies. These institutions are women centred providing loans as well as alternative livelihoods. This has resulted in women empowerment, which builds sustained human capital (see Box 4.6). The impact can be seen in three areas: (i) improvement in living conditions (personal hygiene, clean upkeep of home, better equipped homes, toilets, TVs, etc., adding more self-respect and social status), (ii) skill development and women's place in the family and society (consultative decision-making at home, increased women participation in the CBOs and community affairs), and (iii) participation and development of leadership skills in women (mobilizing govt. programmes, bank credit, representation in PRIs).

While spending on health is a compulsion, spending on education is a household choice. Expenditure on health reflects the health status of the household as well as its affordability. That is, poor households tend to spend on health only when it becomes unavoidable. Across the sample watersheds, households spend on an average Rs. 8000 to Rs. 18,000 per year on health (doctors, medicines, hospitalisation, etc.). Households in the control villages spend less, i.e. Rs. 5500 to Rs. 11,500 per year. Across the economic groups, large farmers tend to spend more on health when compared to small and marginal farmers in all the sample villages, except D. K. Thanda.

On the other hand, household expenditure on education depends on their ability and perceptions about education. Given the fragile nature of household economies

---

Box 4.6 Women Empowerment as Human Capital

While the improvement in living conditions is tangible and measurable, many changes are taking place particularly in the personality of women. They have become more confident and independent. For example, Manjula could not travel alone to buy sarees for her business earlier; her husband used to accompany her. Now, although illiterate, Manjula is travelling on her own without anybody's help. She has learned the skills of bargaining and negotiating with suppliers. Now she is an entrepreneur. Nallamma from Papampalli village has become a role model for other women by becoming a woman auto owner – cum – operator. Sridevi from Battuvanipalli started a petty shop first and expanded to selling fresh vegetables recently. Everybody in the village knows her now. All these women were agricultural labourers once who did not know about the outside world and who belonged to the disadvantaged groups. There is a vast improvement in their skills after they have taken up livelihood activities. Now their voices are heard and decisions are respected in the family as they are contributing to family finances. Like them, many women in the project area have taken up dairy, goat/sheep rearing, non-farm enterprises, etc., to earn additional income.

Source: Personnel Interviews.

in these regions, it is difficult to expect them to spend on education beyond bare minimum. But, households are spending substantial amounts on their children's education. Among the watershed villages, households spend between Rs. 20,000 and Rs. 69,000, while in the control villages, households spend Rs. 15,000 to Rs. 16,000. Higher expenditure on education among the watershed villages is mainly due to their better awareness levels and also higher affordability. In most of these villages, households send their children to private schools in the nearby towns, and engineering is their preferred higher education, which requires higher annual expenditure. An interesting aspect is that relative share of education in the household expenditure is quite high in these villages. Despite high poverty in this region in majority of the sample watersheds, households spend more on education than on food (Figs. 4.2a, 4.2b, 4.2c, 4.2d, 4.2e, 4.2f, and 4.2g). The share of education is substantially less in the control villages, though they spend as much as 20% of their income (see Appendix III Table 4.19A). This indicates very high preference for education in the region in general and in the watershed villages in particular. It is observed that households in this region go to any extent to educate (higher) their children. However, the rationale for such high investment in education is not very clear, as most of the children end up getting poor quality education and settle for low-paying jobs or remain unemployed. This is an interesting revelation, as some of the regions with similar poverty levels spend much less on education (Reddy 2017). The high preference for education coupled with access to credit could be the reason for such behaviour. Most believe that education only can help them come out of distress (See Box 4.7, Table 4.15).

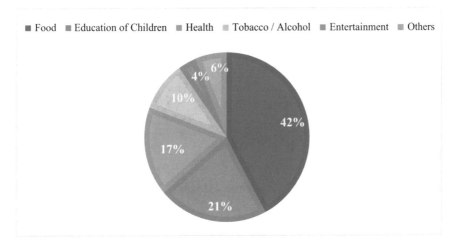

Fig. 4.2a Composition of household expenditure – D. K. Thanda
Source: Field Survey

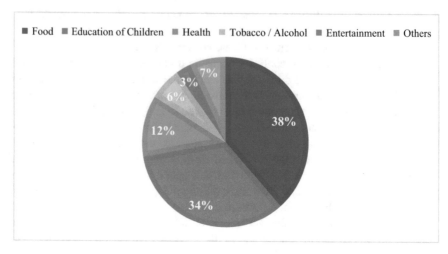

Fig. 4.2b Composition of household expenditure – Goridindla
Source: Field Survey

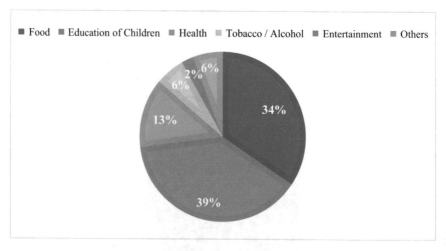

Fig. 4.2c Composition of household expenditure – Muttala
Source: Field Survey

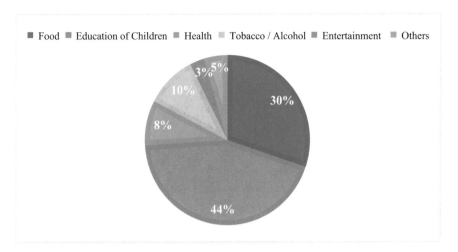

Fig. 4.2d   Composition of household expenditure – Papampalli
Source: Field Survey

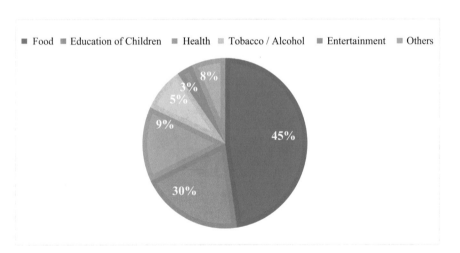

Fig. 4.2e   Composition of household expenditure – Battuvanipalli
Source: Field Survey

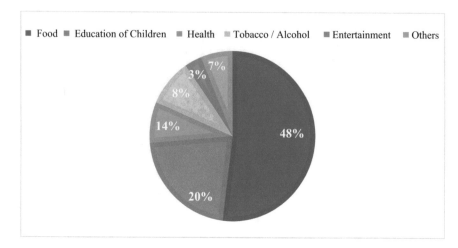

Fig. 4.2f Composition of household expenditure – Control-IWMP
Source: Field Survey

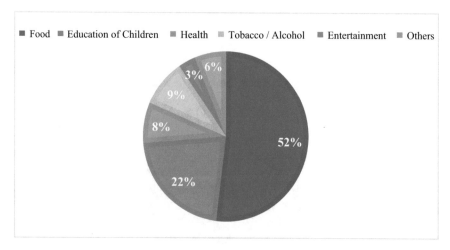

Fig. 4.2g Composition of household expenditure – Control-NABARD
Source: Field Survey

4.3.6   Social Capital

Social capital plays a critical role in evolving institutions and sustaining them, especially in the context of NRM. Social capital is basically assessed in terms of group or collective behaviour which is measured in terms of membership in groups and association with networks that include political, administrative, friends, relatives, etc. The success of watershed interventions is linked to the collective and participatory approaches adopted. Given the importance of community participation in WSM, the programme is often termed as participatory watershed development

Box 4.7 Education as Saviour
*Many families are spending on education of children by borrowing from groups and the additional income from livelihoods. Education of children is now given the highest priority as they see no big hope or future in agriculture. **Varalakshmi** explains the reasons: "An educated person is respected, considered as knowledgeable, and the social status goes up in the society. We don't want our children to work as farmers/labourers like us. This is a hard life. We want our children to have a job, live in the cities and lead a better life than us". Many young men in the project area are studying engineering or basic degrees in other cities. While only some girls are pursuing higher education courses like nursing or teaching, many girls are studying at least until intermediate (12th grade). Parents are often sending their children to English medium schools. Studying in English medium schools is more expensive but perceived to have better employment opportunities than studying in vernacular medium and parents are willing to bear the extra cost.*
Source: Personnel Interviews

Table 4.15 Household expenditure on education and health among the sample households (Rs./household)

Watershed	Small & marginal		Medium		Large		All	
	Education	Health	Education	Health	Education	Health	Education	Health
D. K. Thanda	24,154	12,333	19,700	22,235	1,0800	7800	20,106	15,975
Goridindla	25,167	10,450	52,444	16,875	53,333	15,500	38,917	13,525
Muttala	33,778	9406	69,154	23,833	47,000	24,389	52,828	18,188
Papampalli	27,556	9893	74,111	16,500	110,000	14,300	69,038	13,473
Battuvanipalli	19,000	7658	30,000	11,714	34,286	7643	26,864	8363
Kurlapalli (C-IWMP)	14,000	10,523	14,429	11,731	25,000	15,100	16,238	11,488
N. Puram (C-NABARD)	10,147	5288	26,167	5875	30,000	6500	15,580	5588

Source: Field Survey

programme (PWSDP). Creating watershed institutions like watershed association (WSA), watershed committee (WSC), user groups (UGs), etc. is the starting point in its implementation. Besides, AF-EC has been promoting new groups like SMGs for supporting sustainable agricultural practices and sustaining the watershed interventions and activities. And in the case of NABARD watersheds, MACS are created to sustain the watershed interventions and livelihood activities of the communities. As a result, most of the households in watershed villages become members in one group or the other. Proportion of households having group membership ranges between 84 and 96. Watershed communities (villages) are more active in this regard when compared to the non-watershed (control) villages, where proportion of households having group membership is slightly lower. Though landless households are also eligible to join the relevant groups, their participation is on the lower side when compared to landed households, which is due to the reason that landless often take part only in the credit and livelihood groups. These groups are stronger and more cohesive and help build social capital in these villages.

Though SHGs were popular a decade ago, they were not functional and effective prior to watershed interventions. After the advent of WSDP, these groups have become very active and effective in all the watershed villages. They meet at least once every month. Often meetings are called as per the requirement and meetings conducted business like rather than a formality. For, the livelihood-linked loans are given through these groups. Priority for loans is given to marginal communities like SC/ST/BC/single woman-headed families. Loans are given to women members of SHG for accountability, ease of recovery, and administration. Funds are managed by WDC, of which VO (village federation of SHGs) leaders are also members. Vasantha, a SHG member explains the procedure, *"When a member applies for a loan, VO and WDC carefully evaluate her situation, credit worthiness and the activity she wants to take up before sanctioning the loan. They (IWMP) hold the entire family responsible for repayment of the loan. They explain the loan terms to the family and ask about the livelihood activity the member chooses, so as to make sure that it suits her family situation"*. The maximum amount given under an IWMP loan is Rs.30,000, but there is an option of combining several types of loans (bank loan, *sreenidhi*, watershed, poorest of the poor (PoP) funds for SC/ST, etc.), if the loan requirement is more than Rs.30,000. Once the loan is sanctioned, the member repays the loan in 20 monthly instalments. Every month interest is calculated only on the outstanding loan amount. Loan repayments are remitted to the VO, which takes care of bookkeeping and remitting the funds to banks. SHG leaders and other members attend monthly VO meetings.

The principle of loan sanctioning is similar in NABARD watersheds also, except that here the funds are managed and sanctioned by the MACS committee. MACS consists of both women and men as committee members elected by the *Gram Sabha* (*GS*). Livelihood funds are turned over to MACS for management after the completion of watershed works. Entire watershed village has a right to apply for a loan. The MACS committee carefully evaluates the application before sanctioning the loan. There is more involvement and participation by committee members in deciding the beneficiaries, and preference is given to SC/ST/BC/women. Loans are paid off in ten instalments. All transactions are carried out through bank for transparency and accountability.

4.3.7 Mutually Aided Cooperative Society (MACS): A Success Story in Battuvanipalli

The MACS institution was formed with the members of the watershed villages under MACS act 1990. All eligible families were mobilized to take the membership in the institution, and board of directors (BoD) were elected by the members of the

Table 4.16 Salient features of MACS

Sl. No.	Details	Progress (Phy. /Fin.)
1	Name of the MACS and registration number	AMC/ATP/DCO/2014/4097
2	Eligible families to become members of the MACS	188
3	% of families Enrolled in MACS	100
4	Value of each share certificate given to member families (Rs.)	10, 500
5	**Fund details**	
5.1	LH grant received from NABARD (in Rs.)	17, 56, 100
5.2	All other incomes (in Rs.)	
5.3	Total fund available with MACS at the end of the project	27, 06, 000
6.1	Bank balance in MACS SB account (in Rs.)	4, 72, 525
6.2	Bank balance in MACS fixed deposits (in Rs.)	6, 30, 000
6.3	MACS members availing loans	175
7	Total maintenance fund with WDC MF account (in Rs.)	7, 71, 516

Source: Group Discussions with BoD of MACS, Battuvanipalli and Members

MACS. Further, livelihood (LH) and agriculture productivity enhancement (APE) portfolio of WSC was transferred to MACS account. WSC and MACS members decided to revolve the LH funds (NABARD grant). Share capital deposits through non-redeemable share capital certificates carrying an interest of 6% per annum were issued. General body meetings are organized with all members of MACS including BoD and decided to distribute the loans through MACS institution. Trainings were organized to capacitate the BoD by the resource support organisation (RSO) and the resource persons who are identified by the PMU for the purpose. MACS has a membership of 188 households, i.e. 100% enrolment of the eligible households (Table 4.16). Of which 175 households are availing the loans. By the end of the watershed project, about Rs. 2.7 million were given as loans, which include the LH grant from NABARD and APE funds. It also had about Rs. 1.1 million in fixed deposits and savings accounts and another Rs. 0.77 million of maintenance fund.

Battuvanipalli MACS has emerged as a model cooperative in AP. By March 2018, it has provided loans to a total of 608 members with a total loan amount of Rs. 183 million. It reported more than 80% recovery rate (Table 4.17). And these amounts are growing over the period. Majority of the people (85%) borrowed for the purpose of IGAs like purchasing of *milch* cattle, sheep, and goat, auto works, carpentry, fertilizers business, hotel, mechanic shop, rice business, saree business, stonecutting, tractor works, vegetable business, etc. About 9% of the members borrowed for agriculture, viz. for purchasing of inputs, borewell, motor, and drip material. While 3% have taken the loan for children's education, the remaining took for purchase of two wheelers, house construction, and marriage purposes. This indicates loans are mainly provided for income-generating and productive purposes.

Watershed works need regular maintenance for its long-term stability. While WSC is responsible for maintaining the works during the implementation phase,

Table 4.17 Loan disbursal and recovery of MACS (as on March 2018)

Period	No. of borrowers from MACS	Loan amount (Rs.)	Total recovery amount (Rs.)			Maintenance cost (Rs.)	Savings (Rs.)	Fine amount (Rs.)
			Loan amount	Interest	Total			
March 2015 to Dec 2015	174	3,775,000	2,328,500	212,072	2,540,572	37,750	180,500	8314
Jan 2016 to Dec 2016	216	6,510,000	5,564,000	368,409	5,932,409	65,100	204,500	8815
Jan 2017 to Mar 2018	218	8,020,000	6,872,000	592,133	7,464,133	80,200	255,300	27,860
Total	608	18,305,000	14,764,500	1,172,614	15,937,114	183,050	640,300	44,989

Source: MACS Office, Battuvanipalli

MACS takes over the maintenance after the completion. The records and the accounts of the project were properly maintained in the WSC office at Battuvanipalli indicating transparency in formulation, implementation, execution, and operation of the project. Labour payments were made only after proper scrutiny and inspection of work done by one of the WSC members, and only the required amount is withdrawn from the joint account after preparation and verification of the measurement book and muster roll.

Maintenance fund (MF) was created under the programme for post-project requirements. The corpus of the MF is created from peoples' contribution towards maintenance fund Rs.100 per year. As on date, the contribution of the WS community is accumulated to Rs. 0.48 million at the time of project completion. Grant received in advance from NABARD (Rs. 0.398 million) was deposited in saving account of the bank which gives a return of 12%. So far, an amount of Rs. 90,470 was accrued as interest. The total corpus accumulated was to the tune of Rs. 0.96 million by the end of the programme, which was deposited in a fixed deposit account of the bank and only the interest accrued on it is to be utilized for maintenance. The interest at 12% itself would be Rs. 46,000 per annum. In addition to this, the WS community would be contributing to the MF account, as contributions, which is sufficient to maintain the project in future.

4.3.8 Migration

The fragility of the household economy in this region forces them to migrate in search of livelihoods. In fact, migration is a well-established drought adaptation strategy in this region. Checking migration is considered as one of the social capital-related impacts. While migration is not a serious issue in our sample villages, due to long-standing presence of AF-EC and their interventions, it is widespread in the other watershed and non-watershed villages. Both seasonal and distress migration were prevalent in the IWMP watershed villages. Some families used to migrate even outside the state to Karnataka (Bangalore City) for non-farm works like construction labour. Usually, the head of the household and his wife migrate, while children and old age people stay in the village. It is reported that due to watershed works and activities, such migration is halted. People expressed that promotion of alternative employment opportunities has checked migration. According to some households:

> Our family used to migrate in search of work during off-season. We used to face lot of problems regarding food, shelter, etc., during migration. But now because of watershed works, we are getting work in our own village and leading a peaceful life...

> Because of NREGS we have stopped migrating to other towns and cities as we are now getting enough work within our village throughout the year. From our earnings, we could take up ram lambs rearing apart from making crop investment on our 5 acres of rain-fed land.

Thus, watershed interventions have effectively checked migration in most of villages, and hence it has a positive social impact. For, migration causes lot of psychological distress at the household level. Sometimes, it affects family relationships and bonding in a negative manner.

4.3.9   Overall Changes and Significance

The preceding analysis clearly brings out the positive impacts of watershed and related interventions on all the five capitals. Composite impact of various indicators could be perceived as high, moderate, and low impact on each capital. Based on the quantitative and qualitative assessment of various indicators, except in the case of physical capital, the impacts could be rated as high across all the watershed villages (Table 4.18). The impacts are low to moderate in the control villages. Midstream watersheds of Goridindla and Muttala have performed better when compared to

Table 4.18 Overall impact of watershed interventions on five capitals across sample villages

| Watershed | Capitals | | | | |
	Natural	Physical	Financial	Human	Social
D. K. Thanda (US)	High	Low	High	High	High
Goridindla (MS)	High	Low	High	High	High
Muttala (MS)	High	Low	High	High	High
Papampalli (DS)	High	High	High	High	High
Battuvanipalli	High	Low	High	High	High
Kurlapalli (C-IWMP)	Low	Moderate	Low	Low	Low
N. Puram (C-NABARD)	Low	Moderate	Low	Low	Low

Source: Based on household survey analysis

upstream and downstream watersheds in the case of IWMP. The analysis also indicated that benefit flows are more for small and marginal farmers in terms of crop productivity, access to credit, etc. When enquired about overall impacts, almost 100% of the sample households perceived that there is an increase in the level of income. This perception is more among landless/small and marginal farmers when compared to large farmers (Table 4.19).

In order to assess the statistical validity of the impacts, a paired "t" test was carried out to test the significance of the differences between before and after watershed for important indicators. The summary of the tests is presented in Table 4.20, and the details are presented in the Appendix III (Table 4.20A). The differences turned out significant for majority of the indicators like improvements in irrigation, including groundwater levels, yields of important crops like tomato, and incomes (Table 4.19). Significance of changes is more conspicuous when IWMP micro-watersheds are combined. The differences between Battuvanipalli and its control are not much, which could be due to the spread of impacts beyond the watershed village (demonstration effect).

Often household perceptions are more valued, as the adoption and sustenance of the technologies and techniques are influenced by how households perceive the impact of these interventions. In order to gauge the perceptions on the impact of watershed interventions, sample households were asked whether there is improvement in their income level, annual variation in incomes, and number of sources of income. In an ideal situation, households are expected to say "increase", "decrease", and "increase" to these three queries, respectively, i.e. increase in income level, decrease in annual variation (less vulnerable), and more sources of income (resilience). These perceptions of the sample households are more conspicuous than observed in household data, though the perceptions reemphasize the quantitative assessment. At the watershed village level, majority (above 50%) of the sample households have indicated that their income levels have gone up. In Battuvanipalli 84% of the households feel so, while 50–60% of the households in the IWMP micro-watersheds reported increase in their income (Table 4.20). On the other hand, decline in year-on-year variations in household income was perceived by about 40% of the households in the IWMP micro-watersheds, while only 14% perceive so in the case of Battuvanipalli. And less than 10% of the sample households perceive that their sources of income have gone up after the advent of watershed interventions. Across the economic groups, the differences in perceptions are not much, though highest proportion of landless perceive their incomes have gone up. Unlike the earlier assessments, there is no evidence in the present case about the bias in the benefit flows in favour of large farmers. It is heartening that the benefit flows from watershed interventions have helped in increasing equity. Whereas, the perceptions on all the three indicators are close to zero in the control villages. This clearly brings out the importance and impact of watershed interventions in the minds of the people.

Table 4.19 Statistical significance of important impacts

Indicator	D. K. Thanda	Goridindla	Muttala	Papampalli	IWMP	C-IWMP	Battuvanipalli	C-NABARD
HH income	+*	+*	+*	+*	+*	+*	+*	+*
Area under irrigation								
Kharif	+NS	+NS	+**	+***	+**	-NS	+NS	+NS
Rabi	+NS	+**	+NS	+NS	+**	-**	-NS	+NS
Depth of borewell	+NS	-*	-NS	-*	-**	-NS	-*	-***
Paddy								
Area	+NS	-**	-NS	-**	-NS	-*	-*	-*
Yield	+*	-*	+NS	-*	-NS	-*	-*	-*
Income	+*	-**	+NS	-*	+NS	-**	-*	-*
Groundnut								
Area	+NS	-*	-NS	-*	-*	-**	-NS	-***
Yield	+*	-**	-NS	-*	-NS	+NS	+***	-
Income	+*	-NS	+NS	-**	+**	+***	+**	+**
Red gram								
Area	+NS	+NS	-NS	-NS	+NS	+NS	+NS	+NS
Yield	+NS	+***	+NS	+NS	+**	+***	+NS	+***
Income	+NS	+***	+NS	+**	+**	+***	+**	+**
Tomato								
Area	+*	+***	+**	NS	+**	+*	+**	+*
Yield	+*	+***	+**	+**	+**	+*	+**	+*
Income	+*	+*	+**	+**	+**	+NS	+**	+*

Note: IWMP = all the micro-watershed together; "+" = increase; "−" = decline; NS = not significant; *, **, and *** = significant at 1, 5, and 10% levels.
Source: Based on household survey analysis

Table 4.20 Household perceptions regarding WSD impact on income (% HH)

Watershed/village	Landless			S&M			Medium			Large			All		
	Level	Variation	Sources	Level	Variation	Sources	Level	Variation	Sources	Level	Variation	Sources	Level	Variation	Sources
D. K. Thanda	100	0	0	50	50	6	65	29	6	0	100	0	60	38	4
Goridindla	80	10	0	45	35	20	38	63	0	50	50	0	50	40	8
Muttala	70	20	10	50	44	6	60	40	0	44	44	11	56	38	6
Papampalli	90	10	0	40	53	7	53	47	0	30	50	20	52	42	6
Battuvanipalli	90	10	0	74	26	0	100	0	0	86	7	7	84	14	2
C-IWMP	30	0	0	0	0	0	0	0	0	0	0	0	6	0	0
C-NABARD	0	10	20	0	0	0	0	0	0	0	0	0	0	2	4

Source: Household Survey

4.4   Drought/Climate Resilience of Watershed Communities

Watershed interventions are aimed at strengthening the natural resource base through SMC practices. These interventions are likely to enhance the resilience of the natural resource base and the communities/households depending on it. However, the resilience aspects of watershed impacts are not given due importance in the traditional impact assessments. Despite the widely accepted climate change impacts on dryland agriculture and farmers, resilience is yet to be considered while designing watershed interventions. Perhaps AF-EC is the first organization that realized the importance of enhancing household resilience to drought and climate change and integrated resilience into its WSDP.

As discussed in the earlier chapters, AF-EC has initiated a number of interventions that help enhancing resilience of the communities. Here an attempt is made to assess the resilience of the watershed communities in relation to the five capitals. For this purpose, respondents were asked whether their present assets or abilities help them withstand one, two, or three droughts. A household which is capable of withstanding more than two droughts is considered as more resilient given their capital status. It may be noted that "not resilient" here indicates that household is not able to maintain its normal standard of living or come back to normalcy quickly. For, households survive number of consecutive droughts with declining standards of living. Communities feel that farming continues to be the main livelihood support system, as most of them feel that the chances of moving into non-farm/off-farm livelihoods are not high. Sample households perceive that a combination of natural and physical capital only could enhance their resilience. Given the importance of these two capitals and the ease in their measurement, respondents were asked to indicate additional requirement in natural and physical capital to survive more droughts. In the case of natural capital, irrigated area is used as an indicator, and in the case of physical capital, small and big ruminants are used as indicators. If a household indicates that no additional units of these indicators are required to survive two or three droughts, then the household may be considered as resilient. A household may indicate greater resilience in case of one indicator and lesser resilience in the case of another indicator. In the case of financial, human, and social capitals, respondents were asked how many droughts they can survive given their present status.

As indicated by majority of the farmers, a combination of natural and physical capital is required in order to make them drought resilient. As far as natural capital is concerned, the present level of area under irrigation is barely enough to survive one drought, as households are looking for additional area under irrigation in most of the villages (Table 4.21). On an average, 1 additional acre of irrigation is required to survive one drought, 1–2 acres of additional irrigation is required to survive two droughts, and 2–3 acres are required to survive three droughts. The additional requirement is much higher in the case of physical capital (small as well as big ruminants). This could be due to the very low ownership of livestock consequent to consecutive droughts (see impact section above). A note of caution: not to read too

Table 4.21 Natural and physical capital requirements for household's drought resilience

Watershed	NC (irrigated area in acres)			PC (No. of small ruminants)			PC (no. of big ruminants)		
	1D	2D	3D	1D	2D	3D	1D	2D	3D
D. K. Thanda (US)	0 (2)	1	2	13 (3)	18	25	2 (1)	2	4
Goridindla (MS)	1 (2)	1	2	14 (0)	20	26	1 (1)	2	4
Muttala (MS)	1 (3)	2	2	16 (2)	22	29	2 (1)	3	5
Papampalli (DS)	1 (2)	2	3	13 (6)	20	28	2 (1)	3	4
Battuvanipalli	2 (2)	2	3	12 (1)	23	26	2 (1)	3	4

Note: figures in brackets are present status
Source: Household Survey
NC natural capital, PC physical capital, 1D one drought, 2D two droughts, 3D three droughts

Table 4.22 Drought resilience of the communities with the given financial, human, and social capitals

Watershed	FC			HC			SC		
	1D	2D	3D	1D	2D	3D	1D	2D	3D
D. K. Thanda (US)	94	32	0	66	48	12	78	42	0
Goridindla (MS)	80	32	2	50	38	2	78	34	6
Muttala (MS)	96	40	2	60	42	8	76	44	16
Papampalli (DS)	98	32	4	60	50	8	80	38	6
Battuvanipalli	98	40	4	50	40	6	76	52	6

Source: Based on Household Survey Analysis
FC financial capital, HC human capital, SC social capital, 1D one drought, 2D two droughts, 3D three droughts

much into the suggested figures. They are only indicative of the gap in drought resilience. In the case of other capitals, most of the households indicated that they can manage one drought given their present income status and 60–80% of the households can manage one drought with the given human and social capital status (Table 4.22). Barely any household can manage three droughts with their present capital status. While the process of resilience building is set, it needs to be more widely adopted. While it is clear that access to water is critical for improving resilience of the households, institutional approaches like sharing of groundwater and social regulation to ensure equitable access to the limited groundwater resources are explored in the following chapter.

Appendix

Table 4.1A Impact of WSD on availability of drinking water and changes: IWMP Muttala

Particulars/micro-watershed	Before project	After project
D. K. Thanda MWS		
Open wells (nos.)	–	–
Borewells (nos.)	–	1
Hand pumps (nos.)	2	–
RO plants (nos.)	–	1
Adequacy of drinking water	No	Yes
Quality of drinking	Fluoride	Good
Time spent for etching drinking water (hours)	1	0.5
Depth to water table (in feet)	100	70
Goridindla MWS		
Open wells (Nos.)	–	–
Borewells (Nos.)	1	3
Hand pumps (Nos.)	–	–
RO plants (Nos.)	–	1
Sri Sathya Sai Water Supply Project/Scheme	Yes	Yes
Adequacy of drinking water	No	Yes
Quality of drinking water	Fluoride	Good
Time spent for fetching drinking water (hours)	2	1
Depth to water table (in feet)	150	120
Muttala MWS		
Open wells (nos.)	–	–
Borewells (nos.)	2	4
Hand pumps (nos.)	7	7
RO plants (nos.)	0	1
Sri Sathya Sai Water Supply Project/Scheme	Yes	Yes
Adequacy of drinking water	No	Yes
Quality of drinking water	Fluoride	Good
Time spent for fetching drinking water (hours)	2	1
Depth to water table (in feet)	250	220
Papampalli MWS		
Open wells (nos.)	–	–
Borewells (nos.)	1	1
Hand pumps (nos.)	2	3
RO plants (pos.)	0	1
Sri Sathya Sai Water Supply Project/Scheme	Yes	Yes
Adequacy of drinking water	No	Yes
Quality of drinking water	Fluoride	Good
Time spent for fetching drinking water (hours)	1	0.5
Depth to water table (in feet)	150	140

Providing community managed purified drinking water plants
Source: PRA/FGD Methods and Watershed Records

Table 4.2A Cropping pattern changes in IWMP micro-WS villages: Muttala, Papampalli, Goridindla, and D. K. Thanda

Intervention	Condition	Cropping pattern
Pre-watershed	Irrigated	Majority of the farmers cultivated paddy, groundnut, sunflower, and maize for commercial purposes
		Some of the farmers cultivated food crops of ragi, jowar, and vegetables mostly for their household consumption
		Very few farmers cultivated sweet lime horticulture, but due to water scarcity, declining of groundwater table, and drought situations, borewells are dried up and as a result of that, almost all these trees are dried up
		Some of the farmers also cultivated mulberry/sericulture crop, but due to water scarcity, declining of groundwater table, and drought situations, they stopped this crop/activity
		All these crops are irrigated under flooding method
	Un-Irrigated	Almost all the farmers cultivated groundnut with red gram as the inter-crop for sale
		Some cultivated groundnut with cowpea, groundnut with green gram, and
		groundnut with sesamum and cowpea are the inter-crops
		bajra was the main crop
		For household consumption and the livestock small holders cultivated horse/cow gram, jowar
During watershed	Irrigated	Main crops cultivated during watershed period are vegetables, viz. tomato, green chilli, brinjal, followed by lady's finger, country beans/bottle gourd/cluster beans, bitter guards, and ridge guard
		Nearly 25% of farmers cultivated groundnut under sprinkler method of irrigation;
		Almost 80–90% of farmers cultivated horticultural crops, viz. mango (mono-horticulture) with vegetables, groundnut, pulses as inter crops followed by *sapota* (chiku),
		custard apple, *allaneredu* (Jamun) and others are mixed horticultural crops
		Some of the farmers also cultivated mulberry/sericulture crop
		Initially farmers irrigated their crops under flooding method, and later all are connected to micro-irrigation systems (groundnut under sprinkler and all the other fruits and vegetables under drip)
		Almost all the crops are (annual/horticultural crops) following the MI method of irrigation and also followed the water user's efficiency mechanisms for sustainability of the crop/maintaining of the groundwater table/level
	Un-Irrigated	Almost all the farmers cultivated intercrop viz., red gram with castor; groundnut with red gram; green gram and cowpea; Bengal gram in black soils during *Rabi* season; jowar with cowpea for fodder crops for the cattle
		Bajra crop (as a main crop) was not cultivated during WS period when compared to pre-watershed
		Some farmers also cultivated mango under dry lands (initially watering through manually for survival of the plants).

(continued)

Table 4.2A (continued)

Intervention	Condition	Cropping pattern
At present	Irrigated	Continuing the same crops that are cultivated during the WSDP period
		Almost all the crops are connected to MI systems
		Following the water user's efficiency mechanisms like mulching for vegetable crops during summer/low rainfall and moisture stress situations
	Un-Irrigated	Almost all the farmers are following the inter-cultivation and mixed cultivation methods/practices
		Inter-cultivation crops: red gram with castor and groundnut with red gram
		Mixed cultivation crops: red gram with Jowar/green gram/sesamum/cowpea
		Fodder cultivation crops: jowar with horse/cow gram
Reasons for change in cropping pattern	Irrigated	Impact of WSDP works (SMC/RWHS), requisite awareness, extension, social mobilization, trainings/exposure visits extended by PIA (AF-EC)
		Initially AF-EC is a PIA of WSDP requisite awareness/motivation/trainings/exposure visits extended to the WSC/Sub-committee (user groups); all these are translated to the farmers of the villages
		Later farmers recapture the previous experience/situations towards crop failures towards drought/uneven and in adequate rainfall situations
		MI support by the WSDP, AF, MGNREGS, and other agencies on the basis of convergence
		Cultivating/shifted to vegetable crops tanking in to account the less water, short period, low cost of cultivation, season/area specific, and remunerative
		Choosing or changing of crops influenced by the market situations particularly the vegetable crops
		Paddy crop almost stopped during *Rabi* and summer seasons due to awareness/motivation created by the AF-EC; this was realised/accepted by all the farmers learned from their experience
		Moved towards annual crops to horticultural crops by all the requisite support, extension, and efforts by the PIA
		Reduced the area under groundnut crop due to increased cost of cultivation and lack of remunerative price when compared to vegetable crops (less cost of cultivation, short duration, and remunerative)
		Cropping pattern changed by the local situations (groundnut area declined due to fields is adjoining to the reserve forest (RF) area and frequently wild bears damaging the crop)

(continued)

Table 4.2A (continued)

Intervention	Condition	Cropping pattern
	Un-Irrigated	Farmers are aware of the disadvantages of mono-cropping/ single crop by experience
		Area under groundnut crop declined due to increasing cost of cultivation, prevailing drought situations
		AF is extending support, extension, promoting/demonstrating appropriate suggestions to rain-fed agricultural and sustainable agricultural best methods/ identifying/ recommending and demonstrating best suitable practices according to the local situations
		WSDP PI (AF-EC) encouraging/supporting towards promoting inter-cropping and mixed cropping in reducing the crop loss/drought mitigation and reducing drought resilience by supporting seed/machinery/watering related schemes extended to the SMFs

Source: PRA/FGD methods

Table 4.3A Cropping pattern changes in IWMP WS control village: Kurlapalli

Intervention	Condition	Crops
10 years back	Irrigated	*Household consumption*: ragi, paddy, and vegetables
		Market/sale: sunflower, maize, and groundnut
		Horticulture: very few farmers are having mango orchards, but due to continuous droughts and drying up of their borewells, half of the trees are dried up and as a result are not properly developed resulting in low yields
	Un-Irrigated	*Inter-crops:* groundnut with red gram;
		Single crop: castor for commercial use
		Other crops: bajra, cow gram, jowar are the single crops cultivated for their household consumption and fodder for their cattle
At Present	Irrigated	Up to 5–7% of the farmers are only having irrigation facilities, and they moved from paddy and groundnut crops to vegetable crops (tomato and brinjal)
		This is possible because MI method is being adopted over the past 5–6 years
	Un-Irrigated	Farmers are cultivating in a trial and error basis in growing groundnut with red gram as an inter-crop
		Castor and cowpea are single crops
Reasons for No change/change in cropping pattern	Irrigated	Lack of institutional support, i.e. WSDP works and activities (treatment)
		Drying of up existing borewells
		Lack of water bodies/irrigation sources
	Un-Irrigated	Cultivating less cost of cultivation crops
		Neither WSDP nor any ecology programme by AF-EC covered in this village
		Cultivating Insurance coverage crops and less premium cost crops

Source: PRA/FGD methods

Table 4.4A Cropping pattern changes in NABARD WS village: Battuvanipalli

Intervention	Condition	Crops
Pre-watershed	Irrigated	*For household consumption:* ragi, jowar, paddy, and vegetables
		For market/sale: sunflower and groundnut
	Un-irrigated	*Inter-crops:* groundnut with red gram; groundnut with castor; groundnut with bajra
		Mixed crops: groundnut with cowpea and korra; groundnut with green gram and cowpea
Post-Watershed	Irrigated	**Fruit crops:** cultivated perennial as well as short-duration high value crops viz., mango, sapota, sweet lime, papaya, watermelon, sweet cucumber
		Vegetable crops: tomato, brinjal, country beans, onion, lady's fingers, onion, tapioca, etc.,
		Other crops/annual crop: Limited extent of groundnut
	Un-irrigated	*Inter crops:* groundnut with red gram; groundnut with castor
		Mixed crops: groundnut with green gram, cowpea, and sesamum
		Multiple crops: Navadhanya (nine grains), viz. groundnut, red gram, green gram, bajra, jowar, cowpea, sesamum, korra, black gram
Present	Irrigated	Continuing the same crops during WSDP period
		Following and connected the MI systems with mulching method and other appropriate crop-specific water user's efficiency mechanisms
		Border crops: coconut, agroforestry, and other fruit crops
		Following the crop rotation methods
	Un-irrigated	Continuing the same crops during WSDP
		Very limited number of farmers cultivating with little extent of area under rain-fed groundnut
Reasons for change in cropping pattern	Irrigated	AF-EC WSDP programme works (RWHS, etc.,), activities, awareness, financial support (horticulture with MI etc.), extension, etc.
		Labour scarcity
		Experience in previous crop situations
		Continuous drought and water scarcity situations
		Declining of groundwater and failure of borewells
		Water user's efficiency mechanisms
		Market situations with regard to vegetable crops
		Following the cost benefit calculations
	Un-irrigated	AF-EC WSDP programme works (SMC works, rain-fed and sustainable agricultural farming), activities, awareness, financial support, extension, etc.
		Widespread and continuous drought
		Cultivating low cost and short duration crops
		Prices/market demand
		Experience in previous crop situations
		Continuous drought and Water scarcity situations

Source: PRA/FGD Methods

Table 4.5A Cropping pattern changes in NABARD WS control village: Narayanapuram

Intervention	Condition	Crops
10 Years back	Irrigated	*For household consumption*: ragi, paddy, and vegetables
		For market/sale: groundnut, maize, and sunflower
		Horticulture: very few farmers are having mango orchards, but due to continuous droughts and drying up of borewells, half of the trees are dried up and as a result are not properly developed with low yields
	Un-Irrigated	*Inter-crops:* groundnut with red gram
		Single crop: castor
		Other crops: bajra, cow gram, and jowar are the single crops cultivated mostly for household consumption, seed, and fodder for their cattle
Present	Irrigated	Only 5–6% of the farmers are having irrigation facilities, and they shifted from paddy and groundnut crops to vegetable crops (tomato and brinjal)
		This less extent of irrigated area is possible only because of following the MI method since 5–6 years
	Un-Irrigated	Only 100 farmers are cultivating on a trial and error basis growing groundnut with red gram as an inter-crop
		Castor and cow gram are single crops
Reasons for change in cropping pattern	Irrigated	Lack of institutional support, i.e. WSDP works and activities (treatment)
		Drying up of existing borewells
		Lack of water bodies/irrigation sources
	Un-Irrigated	Cultivating less cost of cultivation crops
		Neither WSDP nor any ecology programme by AF-EC covered in their village
		Mostly cultivating crop insurance coverage/low amount of premium crops

Source: PRA/FGD methods

Table 4.6A Livestock income change of percentage before and after

Watershed	Landless	S&M	Medium	Large	All
D. K. Thanda		−95	−88	−98	−92
Goridindla		−92	−95	−100	−94
Muttala		−90	−90	9	−74
Papampalli	262	97	−50	27	2
Battuvanipalli	−65	−90	−74	−71	−80
Kurlapalli	−67	−38	−68	−70	−63
Narayanapuram		−81	−37	−43	−56

Source: Field Survey

Table 4.7A Cost of livestock rearing change of percentage before and after

Watershed	Landless	S&M	Medium	Large	All
D. K. Thanda	100	−37	−56	−98	−68
Goridindla		−56	−28	−78	−44
Muttala		−50	−38	329	28
Papampalli	97	956	29	285	139
Battuvanipalli	−35	−40	−3	20	−15
Kurlapalli	13	83	8	169	49
Narayanapuram	100	−26	51	−15	17

Source: Field Survey

Table 4.8A Share of incomes before watershed

Sources	Landless							S&M							Medium							Large						
	1	2	3	4	5	C1	C2	1	2	3	4	5	C1	C2	1	2	3	4	5	C1	C2	1	2	3	4	5	C1	C2
Agriculture	0	0	0	0	0	0	0	50	44	35	76	27	27	33	42	70	47	76	61	50	67	74	77	59	79	91	77	67
Livestock	0	0	0	0	0	2	0	7	9	18	0	10	5	6	13	11	19	4	9	2	11	10	3	11	3	7	6	4
Non-farm labour	32	29	49	38	27	38	24	6	3	7	3	6	13	9	4	1	5	1	0	5	0	4	0	6	0	0	0	0
Farm labour	38	57	27	37	46	34	54	21	24	17	14	30	42	37	20	10	10	11	9	16	9	7	7	3	0	1	12	2
NREGA	9	14	18	14	17	4	18	15	17	10	7	12	13	12	15	7	7	8	6	11	8	5	8	3	1	0	4	1
Employment	17	0	0	11	11	22	0	0	2	0	0	14	0	0	0	0	7	0	12	10	0	0	0	0	16	0	0	23
Business	0	0	0	0	0	0	0	0	0	0	0	0	0	0	0	0	0	0	0	0	0	0	0	0	0	0	0	0
hiring out	0	0	0	0	0	0	0	1	1	13	0	1	0	1	5	1	3	1	3	0	2	0	0	18	0	0	0	0
Others	4	0	5	0	0	0	4	0	0	0	0	0	1	3	0	0	0	0	0	0	3	0	6	0	0	0	0	3

Source: Field Survey

Table 4.9A Share of incomes after watershed

Sources	Landless							S&M							Medium							Large						
	1	2	3	4	5	C1	C2	1	2	3	4	5	C1	C2	1	2	3	4	5	C1	C2	1	2	3	4	5	C1	C2
Agriculture	0	0	0	0	0	0	0	52	53	78	28	37	40	54	60	76	71	67	68	38	73	81	35	74	70	75	75	55
Livestock	0	0	0	2	1	1	6	1	2	2	9	4	6	3	0	2	5	8	5	1	9	1	0	12	14	5	6	1
Non-farm Labour	30	42	11	36	18	31	17	12	2	1	9	5	14	9	8	1	4	3	0	11	0	7	0	3	0	0	6	0
Farm labour	43	45	7	17	26	32	50	19	15	6	28	24	31	23	23	10	8	10	6	17	5	6	1	2	0	1	13	2
NREGA	8	11	3	0	9	0	0	8	7	2	2	4	0	0	8	4	3	0	3	0	0	5	1	0	0	0	0	0
Employment	8	0	12	31	4	11	0	6	19	8	1	25	4	5	0	6	9	12	8	29	0	0	16	3	10	19	0	40
Business	8	0	66	14	39	23	17	1	2	2	15	0	0	5	0	0	0	0	8	0	2	0	18	0	0	0	0	0
hiring out	0	0	0	0	0	0	0	0	0	1	10	0	4	0	1	1	1	0	3	4	0	0	0	5	5	0	0	0
Others	4	2	2	1	2	2	11	0	0	0	0	0	1	1	0	0	0	0	0	0	11	0	27	0	0	0	0	2

Source: Field Survey

Table 4.10A Averages of gross income per acre before and after (Rs.)

Watershed	Small & marginal			Medium			Large			All		
	Before	After	% change	Before	After	% change	Before	After	% change	Before	After	% change
D. K. Thanda	13,499	35,091	160	11,946	27,799	133	22,216	24,869	12	14,182	30,010	112
Goridindla	10,081	30,113	199	26,871	43,746	63	6192	23,542	280	17,574	35,290	101
Muttala	9689	68,291	605	12,049	38,046	216	17,306	34,628	100	12,658	46,826	270
Papampalli	13,347	28,488	113	19,315	38,292	98	23,051	44,073	91	19,310	38,292	98
Battuvanipalli	7374	29,597	301	11,501	31,891	177	12,930	27,398	112	11,049	28,968	162
Kurlapalli (C-IWMP)	6722	20,472	205	13,624	18,907	39	27,625	26,764	−3	13,848	21,397	55
N. Puram(C-NABARD)	8296	34,389	315	10,365	33,862	227	11,578	36,174	212	9644	34,552	258

Source: Field Survey

Table 4.11A Averages of net income per acre before and after (Rs.)

Watershed	Small & marginal			Medium			Large			All		
	Before	After	% change	Before	After	% change	Before	After	% change	Before	After	% change
D. K. Thanda	6067	20,440	237	4841	15,345	217	12,554	14,702	17	6545	17,148	162
Goridindla	5035	17,877	255	13,470	25,090	86	2746	15,221	454	8759	20,717	137
Muttala	4851	36,169	646	5219	23,303	347	7815	24,268	211	5804	27,711	377
Papampalli	6417	17,222	168	9153	21,663	137	11,561	28,018	142	9402	23,064	145
Battuvanipalli	3611	17,597	387	6137	18,107	195	7028	13,855	97	5869	15,815	169
Kurlapalli (C-IWMP)	2452	15,491	532	7069	10,877	54	12,387	14,366	16	6453	13,469	109
N. Puram(C-NABARD)	4110	21,521	424	5176	21,391	313	6091	22,418	268	4876	21,644	344

Source: Field Survey

Table 4.12A Crop wise changes in yields (value and cost of cultivation different indicators of sample households by economic groups (% change before/present)

Crop: groundnut

Watershed	S&M				Medium				Large				All			
	OV	IC	NI	Y	OV	IC	NI	Y	OV	IC	NI	Y	OV	IC	NI	Y
D. K. Thanda	128	56	276	26	83	75	212	46	19	2	45	0	84	51	185	29
Goridindla	33	42	1	28	133	130	227	39	-16	-20	22	-22	83	82	94	30
Muttala	82	-3	182	32	136	82	159	98	96	-35	251	66	102	23	173	62
Papampalli	104	170	107	19	48	52	95	19	24	-9	88	11	49	45	97	17
Battuvanipalli	122	114	220	32	116	142	137	34	70	72	107	46	109	112	164	49
Kurlapalli (C-IWMP)	266	362	282	123	105	87	167	44	46	98	22	15	150	148	219	68
N. Puram (C-NABARD)	128	87	242	38	137	43	303	62	41	0	63	-11	110	50	199	28

Source: Field Survey

Note: OV output value, IC input costs, NI net income, Y yield

Crop: red gram

Watershed	S&M				Medium				Large				All			
	OV	IC	NI	Y	OV	IC	NI	Y	OV	IC	NI	Y	OV	IC	NI	Y
D. K. Thanda	22	54	41	33	66	56	137	36	6	0	8	-13	25	37	38	21
Goridindla	148	71	192	65	125	171	161	111	40	9	13	25	115	80	94	67
Muttala	18	15	43	37	9	15	15	-1	18	-1	106	11	15	14	46	15
Papampalli	104	92	216	39	38	77	62	11	74	38	176	45	71	68	147	33
Battuvanipalli	94	15	105	4	62	31	136	78	38	41	35	14	54	32	61	20
Kurlapalli (C-IWMP)	17	35	94	11	24	31	26	1	141	174	264	86	33	49	58	13
N. Puram (C-NABARD)	96	114	141	36	30	49	48	9	37	79	72	24	66	89	100	26

Source: Field Survey

Note: OV output value, IC input costs, NI net income, Y yield

Crop: tomato

Watershed	S&M				Medium				Large				All			
	OV	IC	NI	Y	OV	IC	NI	Y	OV	IC	NI	Y	OV	IC	NI	Y
D. K. Thanda	119	91	174	22	127	115	174	26	-88	67	-145	40	85	93	100	23
Goridindla	106	80	169	33	6	17	24	17	100	100	100	100	20	26	42	14
Muttala	158	70	241	187	40	-14	113	20	16	-43	109	35	50	-11	137	55
Papampalli	76	35	146	44	72	77	83	29	4	-17	30	11	49	36	76	27
Battuvanipalli	179	273	169	40	140	227	123	47	-18	53	-6	27	61	154	67	34
Kurlapalli (C-IWMP)	100	100	100	100	28	50	24	5	-47	0	-31	-13	27	16	83	-2
N. Puram (C-NABARD)	100	100	100	100	168	136	228	120	100	100	100	100	181	149	242	47

Source: Field Survey
Note: *OV* output value, *IC* input costs, *NI* net income, *Y* yield

Crop: castor

Watershed	S&M				Medium				Large				All			
	OV	IC	NI	Y	OV	IC	NI	Y	OV	IC	NI	Y	OV	IC	NI	Y
D. K. Thanda	62	-4	197	25	147	79	398	47	106	52	438	39	106	38	328	36
Goridindla	23	25	-93	-9	41	56	75	13	28	7	89	-5	29	32	-82	-1
Muttala	43	37	137	-1	-57	-65	7	-63	-100	-100	-100	-100	23	4	130	-13
Papampalli	60	28	172	24	16	41	31	3	100	100	100	100	55	59	109	25
Battuvanipalli	-100	-100	-100	-100	0	0	0	0	19	4	76	-19	29	12	94	-14
Kurlapalli (C-IWMP)	-5	4	24	-16	20	-6	68	-18	20	18	89	-6	6	1	44	-17
N. Puram (C-NABARD)	-100	-100	-100	-100	-100	-100	-100	-100	-100	-100	-100	-100	-100	-100	-100	-100

Source: Field Survey
Note: *OV* output value, *IC* input costs, *NI* net income, *Y* yield

Table 4.13A Averages of savings before (Rs./HH)

Watershed	Landless	S&M	Medium	Large	All
D. K. Thanda	367	574	1137	433	710
Goridindla	617	1275	875	5417	1347
Muttala	267	1573	1244	2333	1350
Papampalli	1017	1022	662	1000	909
Battuvanipalli	1617	1167	571	6238	2593
Kurlapalli	1000	379	769	800	647
Narayanapuram	583	987	792	5000	1357

Source: Field Survey

Table 4.14A Averages of savings after (Rs./HH)

Watershed	Landless	S&M	Medium	Large	All
D. K. Thanda	1617	2778	3578	1733	2713
Goridindla	1583	6767	5577	6750	5348
Muttala	817	7188	8633	7130	6337
Papampalli	2033	3511	3556	10,100	4547
Battuvanipalli	2767	7228	7238	20,500	10,053
Kurlapalli	2083	2659	7833	2933	3917
Narayanapuram	1700	4359	4708	13,833	5020

Source: Field Survey

Table 4.15A Change of percentage of savings before and after (% change)

Watershed	Landless	S&M	Medium	Large	All
D. K. Thanda	341	384	215	300	282
Goridindla	157	431	537	25	297
Muttala	206	357	594	206	369
Papampalli	100	243	437	910	400
Battuvanipalli	71	520	1167	229	288
Kurlapalli	108	602	918	267	506
Narayanapuram	191	342	495	177	270

Source: Field Survey

Table 4.16A Averages of debts before (Rs./HH)

Watershed	Landless	S&M	Medium	Large	All
D. K. Thanda	0	2667	5529	0	2840
Goridindla	2500	5150	6750	0	4720
Muttala	0	13,750	19,733	0	10,320
Papampalli	2000	3667	15,000	0	6000
Battuvanipalli	28,000	7842	15,714	4071	11,920
Kurlapalli	17,500	4545	5385	12,000	8100
Narayanapuram	11,000	2038	1250	0	3460

Source: Field Survey

Table 4.17A Averages of debts after (Rs./HH)

Watershed	Landless	S&M	Medium	Large	All
D. K. Thanda	30,000	125,556	246,471	254,000	160,400
Goridindla	34,300	152,250	218,063	176,750	151,680
Muttala	26,700	244,313	261,920	320,844	219,848
Papampalli	59,000	114,333	191,787	323,500	168,336
Battuvanipalli	130,300	81,816	181,429	196,071	137,450
Kurlapalli	100,300	119,500	183,462	319,600	152,300
Narayanapuram	50,500	66,615	117,875	193,333	86,800

Source: Field Survey

Table 4.18A Change in percentage of debt before and after (% change)

Watershed	Landless	S&M	Medium	Large	All
D. K. Thanda	100	4608	4357	100	5548
Goridindla	1272	2856	3131	100	3114
Muttala	100	1677	1227	100	2030
Papampalli	2850	3018	1179	100	2706
Battuvanipalli	365	943	1055	4716	1053
Kurlapalli	473	2529	3307	2563	1780
Narayanapuram	359	3168	9330	100	2409

Source: Field Survey

Table 4.19A Consumption choices: money spent (in RS. Average/HH)

Farm Size	Consumption Type	D. K. Thanda	Goridindla	Muttala	Papampalli	Battuvanipalli	C-IWMP	C- NABARD
Small & marginal	Food	35,500	39,450	36,063	37,800	35,842	32,864	32,462
	Education of children	24,154	25,167	33,778	275,56	19,000	14,000	10,147
	Health	12,333	10,450	9406	9893	7658	10,523	5288
	Tobacco/alcohol	7235	6065	5917	10,154	3688	5967	8118
	Entertainment	3276	2563	2213	2164	2147	2545	1962
	Others	5100	5746	6385	4643	4132	3786	3269
	Borewells and motors	5050	4667	8000	11,000	4667	26,000	3050
	Tractor	4000	–	–	4000	–	250,000	–
	Other Implements	1000	–	2000	–	–	200,000	–
	Land development	3056	2250	2063	447	847	2191	496
Medium	Food	42,447	47,500	43,933	46,667	48,000	44,077	45,000
	Education of children	19,700	52,444	69,154	74,111	30,000	14,429	26,167
	Health	22,235	16,875	23,833	16,500	11,714	11,731	5875
	Tobacco/alcohol	10,941	8500	9964	25,750	2333	7727	3500
	Entertainment	3163	2640	3486	3636	2267	2408	2275
	Others	5636	11,278	8955	5150	7357	7250	5625
	Borewells and motors	6188	5688	8500	14,571	6200	25,500	6083
	Tractor		4000				70,000	
	Other Implements			6000				
	Land development	2353	4450	15,800	2487	1357	2615	313

Large	Food	43,600	56,000	71,778	62,500	42,643	48,600	47,500
	Education of children	10,800	53,333	47,000	110,000	34,286	25,000	30,000
	Health	7800	15,500	24,389	14,300	7643	15,100	6500
	Tobacco/alcohol	7600	6500	6563	7667	4864	6750	3750
	Entertainment	4000	10,250	4150	8500	3243	2680	4000
	Others	10,000		10,750	13,833	10,429	5750	8167
	Borewells and motors	3000	33,000	27,667	18,750	6000	10,000	5000
	Tractor			140,833	16,500			
	Other Implements			40,000	10,000			
	Land development	8000	3750	30,611	6300	1643	1200	1167
All	Food	39,465	44,325	47,050	47,300	40,350	38,475	37,225
	Education of children	20,106	38,917	52,828	69,038	26,864	16,238	15,580
	Health	15,975	13,525	18,188	13,473	8363	11,488	5588
	Tobacco/alcohol	8897	7165	7735	15,710	4091	6717	6444
	Entertainment	3305	3403	3089	4374	2559	2516	2383
	Others	5783	8009	8018	7261	6900	5304	4475
	Borewells and motors	5300	8079	11,647	15,833	5842	21,708	4350
	Tractor	4000	–000	140,833	12,333	–	160,000	–
	Other Implements	1000	–	22,000	10,000	–	200,000	–
	Land development	3375	3280	13,638	2675	1215	2205	560

Source: Field Survey

Table 4.20A Results of Paired "t" test – IWMP all villages: D. K. Thanda, Goridindla, Muttala, Papampalli, NABARD watershed Battuvanipalli and the control villages

Variables	Before	After	Difference	# of observations	Significance
Incomes (Rs./HH)	26,825	70,368	43,543.9	200.0	.000
Irrigation					
Kharif (acres)	1.3	2.7	1.4	160.0	.000
Rabi (acres)	0.6	1.6	1.0	160.0	.000
Depth of borewell (feet)	135.3	117.5	−17.9	47.0	.007
Paddy					
Area (acres)	2.0	1.5	−0.5	30.0	0.173
Yield (qnt/acre)	17.38	15.48	−1.9	30.0	.286
Income (Rs./acre)	9947.9	12,072.2	2124.3	30.0	0.281
Groundnut					
Area (acres)	4.4	2.6	−1.8	109.0	.000
Yield (qnt/acre)	3.5	3.3	−0.3	109.0	.344
Income (Rs./acre)	4028.0	6829.7	2801.7	109.0	.001
Red gram					
Area (acres)	4.0	4.0	0.05	61.0	0.921
Yield (qnt/acre)	2.3	2.9	0.6	61.0	.000
Income (Rs./acre)	2998.1	5305.9	2307.8	61.0	.000
Tomato					
Area (acres)	0.7	1.9	1.2	103	.000
Yield (qnt/acre)	28.7	95.1	66.4	103	.000
Income (Rs./acre)	9907.6	43723.0	33,815.4	103	.000

Source: Based on Household Survey Analysis

Village: Muttala					
Variables	Before	After	Difference	No. of observations	Significance
Incomes(Rs./HH)	3842	85,125	81,283.0	50.0	.000
Irrigation: area					
Kharif	3.7	7.6	3.9	13	.050
Rabi	2.4	4.1	1.8	8	.144
Depth of borewell	135.5	116.9	−18.5	11	.411
Paddy					
Area	2.5	2.0	−0.5	2.0	0.5
Yield	15.16	20.50	5.34	2.00	.263
Income	10,666.7	16,250.0	5583.3	2.0	0.402
Groundnut					
Area	3.7	2.8	−0.89	28.0	0.427
Yield	2.8	2.7	−0.13	28.0	.876
Income	3416.4	5518.8	2102.34	28.0	.236
Red gram					
Area	3.1	2.9	−0.2	29.0	0.738
Yield	2.1	2.2	0.0	29.0	.988

(continued)

Table 4.20A (continueed)

Village: Muttala

Variables	Before	After	Difference	No. of observations	Significance
Income	2858.8	3617.9	759.1	29.0	.444
Tomato					
Area	0.4	2.2	1.7	27.0	.000
Yield	13.3	93.4	80.1	27.0	.000
Income	4446.9	47,458.8	43,011.9	27.0	.000

Source: Field Survey

Village: Papampalli

Variables	Before	After	Difference	No. of observations	Significance
Incomes(Rs./HH)	25,588	69,042	43,454.0	50.0	.000
Irrigation: area					
Kharif	3.6	3.8	0.2	21.0	.104
Rabi	2.9	3.5	0.5	11.0	.367
Depth of borewell	131.5	114.1	−17.4	17.0	.002
Paddy					
Area	1.9	0.4	−1.4	7.0	0.094
Yield	18.92	5.71	−13.2	7.0	.023
Income	10,521.4	2071.4	−8450.0	7.0	0.011
Groundnut					
Area	4.4	1.1	−3.3	24.0	.000
Yield	3.4	0.9	−2.5	24.0	.000
Income	3724.8	1594.2	−2130.6	24.0	.045
Red gram					
Area	3.4	3.1	−0.3	28.0	`
Yield	1.8	2.1	0.3	28.0	.525
Income	2071.3	4424.7	2353.4	28.0	.009
Tomato					
Area	1	1	0.0	24.0	1
Yield	56.6	95.9	39.3	24.0	0
Income	18,326.4	43,100.1	24,773.7	24.0	.000

Source: Field Survey

Village: Goridindla

Variables	Before	After	Difference	No. of Observations	Significance
Incomes(Rs./HH)	19,530	69,487	49,957.0	50.0	.000
Irrigation: area					
Kharif	2.9	4.9	2.0	14.0	.199
Rabi	2.2	2.8	0.6	10.0	.051
Depth of borewell	154.4	123.3	−31.1	9.0	.000

(continued)

Table 4.20A   (continueed)

Village: Goridindla

Variables	Before	After	Difference	No. of Observations	Significance
Paddy					
Area	2.7	0.0	−2.7	3.0	0.057
Yield	18.50	0.00	−18.50	3.0	.009
Income	9833.3	0.0	−9833.3	3.0	0.081
Groundnut					
Area	4.4	1.5	−2.9	32.0	.000
Yield	3.5	2.1	−1.4	32.0	.025
Income	4136.6	3596.9	−539.7	32.0	.721
Red gram					
Area	1.7	2.4	0.7	18.0	0.492
Yield	0.8	1.9	1.1	18.0	.044
Income	1043.5	3042.8	1999.3	18.0	.065
Tomato					
Area	0.58	4.61	4.0	31.0	0.101
Yield	23.9	184.9	161.1	31.0	0.073
Income	10,741.0	44,079.6	33,338.5	31.0	.000

Source: Field Survey

Village: D. K. Thanda

Variables	Before	After	Difference	No. of observations	Significance
Incomes(Rs./HH)	23,838	57,819	33,981.2	50.0	.000
Irrigation: area					
Kharif	2.2	2.9	0.6	21.0	.242
Rabi	1.5	2.2	0.7	12.0	.136
Depth of borewell	124.5	118.5	−6.0	10.0	.714
Paddy					
Area	1.4	1.8	0.4	25.0	0.44
Yield	12.12	19.54	7.42	25.0	.000
Income	6958.1	16,211.6	9253.47	25.0	.000
Groundnut					
Area	3.9	3.2	−0.64	33.0	0.171
Yield	3.5	4.5	1.02	33.0	.014
Income	3685.8	10,498.1	6812.39	33.0	.000
Red gram					
Area	1.5	1.9	0.38	16.0	0.633
Yield	1.0	1.6	0.53	16.0	.367
Income	1450.0	2505.0	1055.03	16.0	.296
Tomato					
Area	0.7	1.4	0.71	21.0	0.003
Yield	32.1	90.3	58.20	21.0	.000
Income	9535.7	39,105.6	29,569.84	21.0	.001

Source: Field Survey

(continued)

Table 4.20A (continueed)

Village: Battuvanipalli					
Variables	Before	After	Difference	No. of observations	Significance
Incomes(Rs./HH)	25,000	83,307	58,307.0	50.0	.000
Irrigation: area					
Kharif	6.6	6.8	0.3	12.0	.191
Rabi	4.5	4.1	−0.4	10.0	.613
Depth of borewell	92.0	72.0	−20.0	5.0	.003
Paddy					
Area	4.3	0.1	−4.2	9.0	0.003
Yield	21.18	2.30	−18.88	9.0	.000
Income	141,090.2	2200.0	−138,890.2	9.0	0.004
Groundnut					
Area	4.3	3.8	−0.5	33.0	0.332
Yield	3.5	4.8	1.3	33.0	.035
Income	3831.8	9183.0	5351.2	33.0	.000
Red gram					
Area	7.4	7.9	0.4	20.0	0.798
Yield	2.5	3.0	0.5	20.0	.153
Income	3938.6	6327.4	2388.9	20.0	.026
Tomato					
Area	0.4	2.5	2.1	34.0	.000
Yield	9.4	85.5	76.1	34.0	.000
Income	2794.6	29,928.7	27,134.1	34.0	.000

Source: Field Survey

Village: Kurlapalli (Control-IWMP)					
Variables	Before	After	Difference	No. of observations	Significance
Incomes(Rs./HH)	32,472	72,250	39,778.0	50.0	.000
Irrigation: area					
Kharif	3.3	3.2	−0.2	6.0	.363
Rabi	3.8	2.3	−1.5	4.0	.058
Depth of borewell	103.3	96.7	−6.7	6.0	.363
Paddy					
Area	2.0	0.4	−1.6	11.0	0.002
Yield	20.57	3.09	−17.48	11.00	.000
Income	10,721.2	2959.1	−7762.1	11.0	0.009
Groundnut					
Area	3.6	2.4	−1.2	18.0	0.098
Yield	2.5	3.0	0.5	18.0	.385
Income	2337.6	5391.7	3054.1	18.0	.030

(continued)

Table 4.20A (continueed)

Village: Kurlapalli (Control-IWMP)

Variables	Before	After	Difference	No. of observations	Significance
Red gram					
Area	2.4	2.9	0.5	23.0	0.492
Yield	1.4	2.5	1.0	23.0	.017
Income	1843.8	4474.8	2630.9	23.0	.003
Tomato					
Area	0.3	1.6	1.3	12.0	.000
Yield	17.1	70.8	53.8	12.0	0.003
Income	10,500.0	55,962.5	45,462.5	12.0	.133

Source: Field Survey

Village: N. Puram (Control-NABARD)

Variables	Before	After	Difference	No. of observations	Significance
Incomes(Rs./HH)	29,394	71,712	42,318.0	50.0	.000
Irrigation: area					
Kharif	4.3	4.6	0.3	11.0	.192
Rabi	4.3	4.8	0.5	11.0	.138
Depth of borewell	103.6	98.6	−5.0	11.0	.067
Paddy					
Area	3.1	0.3	−2.8	12.0	.000
Yield	18.11	2.08	−16.0	12.00	.000
Income	10,362.1	2035.5	−8326.6	12.0	.000
Groundnut					
Area	3.2	1.9	−1.3	28.0	0.017
Yield	3.5	3.4	−0.1	28.0	.847
Income	399.4	8973.1	8573.7	28.0	.000
Red gram					6
Area	3.8	4.0	0.2	26.0	0.457
Yield	2.5	3.1	0.6	26.0	.063
Income	3446.0	6763.9	3317.9	26.0	.000
Tomato					
Area	0.1	2.3	2.3	36.0	.000
Yield	2.1	106.3	104.2	36.0	.000
Income	319.4	38,173.8	37,854.4	36.0	.000

Source: Field Survey

Chapter 5
Mitigating Climate/Drought Risks: Role of Groundwater Collectivization in Arid/ Semi-Arid Conditions

5.1 Background

Access to water (groundwater) in the study villages is critical for sustainable livelihoods and reducing vulnerabilities at the household level. Sustainable groundwater management (SGWM) is critical for food security and poverty alleviation (Shah 2004). Groundwater irrigation is twice as efficient[1] as surface water irrigation in hydrological terms (m³/ha) (Llamas and Martínez-Santos 2005) and requires relatively smaller investment and shorter implementation periods when compared to surface irrigation systems (Valencia Statement 2004). Besides, it has a large number of inherent services including environmental services (Burke et al. 1999; Polak 2004). These virtues of groundwater in the absence of clearly defined property rights[2] have resulted in the sharp increase in groundwater use and over-exploitation (OE) as well as degradation of the resource (Moench 1992; Bhatia 1992; Dhawan 1995). More importantly, there is a clear policy neglect of this important resource over the years. While surface water systems fall under the purview of public policy, groundwater systems are left to the purview of private individuals (Shah 2009).

In the absence of any control or regulation, groundwater has become one of most mismanaged resources. The spread of this uncontrolled exploitation has grown beyond public policy management in countries like India where the number of private bore wells is growing beyond limits, which is termed as "anarchy" (Shah 2009). This is more so in the hard-rock areas like Anantapuramu district with low rainfall

---

[1] Here groundwater irrigation efficiency is defined in terms of reduced non-beneficial and nonrecoverable fractions of water withdrawals as elaborated in Perry (2007).

[2] Since property rights on groundwater are linked to landownership, only landowners benefit from these rights. The landless and even the landowners who are not capable of investing in exploiting groundwater tend to lose. The heterogeneity in the ownership of land as well as financial capabilities result in differences in the perception of the nature of property rights (i.e. private individual based or community based). Hence, there is no agreement on how to design most effective property rights for groundwater (Libecap 1997; Reddy 1999).

© Springer Nature Switzerland AG 2020
V. R. Reddy et al., *Climate-Drought Resilience in Extreme Environments*,
https://doi.org/10.1007/978-3-030-45889-8_5

and limited surface water resources. Number of bore wells and their depths are continuously on the rise due to ever-increasing demand for water. Between 2006–2007 and 2013–2014, the number of deep tubewells wells in the district has gone up from 19,711 to 78,731 (GoAP 2015). During the same period, dugwells declined from 38,616 to 18,290. This clearly indicates the increasing tress on groundwater resources. The first victims of the increased dependence on deep tube-wells wells are small and marginal farmers who can't afford to invest capital on deepening the wells (Reddy 2004, 2005). Greater incidence of climate/drought risks in the recent years has further deepened the groundwater crisis in the region. Declining access to groundwater affects not only agricultural production but also equity, education, health, gender, child mortality, poverty, and hunger (Sharma 2009).

Sustainable management of water is encouraged through institutional arrangements such as water user associations (WUAs) and tank management committees (TMCs). These state-promoted institutional arrangements are limited to surface water resources such as canals and tanks, leaving groundwater development and management out of the public purview. Though the effectiveness and sustainability of canal and tank management institutions are being debated (Reddy and Reddy 2005), the need for bringing groundwater under a common resource management regime cannot be underestimated. Hitherto groundwater management has been left to private individuals as it is perceived to have high transaction costs of organizing individual farmers at a scale to attain the benefits of community management.

As observed in the literature, there appear to be some small-scale institutional innovations that are working towards sustainable management of groundwater in different corners of India, including Anantapuramu district (for a detailed review, see Reddy et al. 2014; Verma et al. 2012). Of late, AF-EC has also initiated groundwater collectivization in some of their villages. While these initiatives are considered as a positive change in groundwater management and policymakers are trying to scale up, whether such initiatives in arid (<500 mm annual rainfall) and semi-arid (500–750 mm annual rainfall) conditions with limited surface water resources would sustain is a moot point. For instance, the need for SGWM is more in these conditions, and policy can help in overcoming the adverse environmental conditions. In this chapter, the institutional modalities of groundwater collectivization and their effectiveness (in terms of sustainability and equity) in arid and semi-arid regions of Anantapuramu district are presented. While the AF-EC initiatives of groundwater collectivization in recent years represent the arid conditions, some of the earlier interventions by other NGOs in the district have been in the semi-arid region. A comparative assessment of these two regions helps to understand the potential and constraints of participatory groundwater institutions in harsh environments for scaling up in the future.

5.2   Groundwater Collectivization: AF-EC Interventions (Arid Conditions)

Between 2012 and 2015, AF-EC has initiated groundwater collectivization with the support of the Department of Agriculture (DoA) and Revitalizing Rainfed Agriculture (RRA: a network) to address the drought distress of the rain-fed farmers. Under this initiative five groundwater collectives were initiated in four villages (Table 5.1). Under this a total of 25 bore wells belonging to 24 farmers were formed into groups of 4–6 bore wells each. These five groups support a total of 55 farmers covering an area of 241 acres. That is, the collectives provide protective irrigation to 32 rain-fed farmers covering an area of 180.5 acres.

5.2.1   Institutional Evolution and Setup

Drawing from the experiences of the Andhra Pradesh Drought Adaptation Initiative (APDAI) implemented by another NGO (WASSAN), AF-EC has evolved a method of sharing groundwater among the farmers within the village. As a first step, the concept and method of collectivization were shared and discussed in the meetings organized with the farmers. Various issues, including the benefits for both bore well-owning farmers and non-borewell farmers, were discussed. Follow-up meetings were organized with farmers to motivate them to come forward and join the proposed initiative. In order to make the farmers join the initiative, AF-EC has

Table 5.1 Details of groundwater collectivization under AF-EC

Sl. No.	Village/ group	Mandal	Avg. rainfall (mm)	No. of bore wells under sharing	No. of farmers			Area (acres)		
					Irrigated	Rain-fed	Total	Irrigated	Rain-fed	Total
1	Kalagalla	Kuderu	347	6	6	7	13	19	39	58
2	M.M. Halli	Kuderu	347	4	5	5	9	10.5	40.5	51
3. A	Korrakodu: A	Kuderu	347	5	5	7	12	14.5	30.5	45
3. B	Korrakodu: B	Kuderu	347	5	4	5	9	6	36	42
4	Sanapa	Atmakuru	341	5	4	8	12	10.5	34.5	45
Total				**25**	**24**	**32**	**55**	**60.5**	**180.5**	**241**

Source: Directors Office, AF-EC, Ananthapuramu

Photo 5.1 Participatory groundwater management (group meeting, measurement, crop water budgeting discussion and supply of water through pipelines)

organized a series of activities (Photo 5.1). These include (i) a survey to estimate the number of bore wells in the village; (ii) GPS survey to assess the groundwater table; (iii) train farmers to measure the water levels in the dugwells and water discharge rates per minute from the bore wells; (iv) linkages between rainfall and groundwater recharge; and (v) exposure visits to the areas where groundwater sharing institutions are functioning and interaction with farmers was organized. All these activities help increasing the awareness and capacities of the communities regarding groundwater status in their village.

As farming is the mainstay of the people in the village, farmers are motivated to think about it in the broader context of promoting the wellbeing of the farming community. Bore well-owning farmers were made aware that they can help their fellow farmers protect their crop by providing irrigation during long dry spells. The system of groundwater sharing is based on certain norms, evolved and agreed by the farmers concerned. It helps in irrigating more area with same water by adopting water-use efficiency mechanisms. Farmers with bore wells and without bore wells in particular area are formed into groups with a purpose to share groundwater between the bore well-owning farmers and those who are not having bore wells, which is termed as area-based approach.

The aims and objectives include:

(i) Adopting an area-based approach for irrigation
(ii) Treating groundwater as a common property
(iii) Checking competitive digging of bore wells
(iv) Providing access to the groundwater for rain-fed crops for protective/critical irrigation
(v) Reducing water loss by adopting effective irrigation systems and methods
(vi) Reducing the cultivation of water-intensive crops (paddy) under bore wells and motivating the farmers for alternative crops to improve water productivity
(vii) Enabling village-level institutions for groundwater regulation, including monitoring of yields of bore wells
(viii) Improving the groundwater recharge
(ix) Determining the extent of land to be cultivated by estimating the groundwater availability in the bore wells (crop water budgeting)

Area-based approach involves organizing farmers under common interest groups (CIGs) for a rain-fed patch. In each patch, well owners were convinced of the efficacy of sharing their water with the neighbouring farmers. Once consensus was reached on water regulations and sharing the cost of installation of the pipeline, an agreement on groundwater regulation was signed by all the farmers in the patch in the presence of a revenue official at the sub-district level on a Rs.100 (US$2) bond paper (Photo 5.2). As per the agreement, all the bore wells will be pooled through a common pipeline network, and water will be shared among all, irrespective of

Photo 5.2 Signing of MoU with the Mandal Revenue Officer

ownership. The bore wells of the farmers willing to share are interconnected to one main pipeline, which is distributed to the identified rain-fed patch of land. No new bore wells will be dug for at least the next 10 years. The cropping pattern will be decided on the basis of crop plans linked to the availability of water in agreement with members of the CIG while giving priority to food and fodder crops. One bore well a day will be rested on rotation, thus reducing water pumping by about 20%, while water is shared to protect the *Kharif* (first crop: June to October) crop of non-owners and the acreage of bore well-owning farmers are ensured. A general fund is created by collecting Rs. 2,500 per acre as membership fee, which is used for the maintenance of the pipeline, repairs, etc. within the CIG.

As there was no threat of new bore wells coming up in the vicinity that may lead to the drying up of their own bore well, farmers agree to pool their bore wells and share the water. The bore well owners are allowed to continue their earlier cropped area under irrigation but with less water-intensive crops. The water thus saved will provide critical irrigation to a rain-fed patch, which includes lands of both owners and non-owners of bore wells. If any one of the bore wells fails, there is a back-up arrangement as they are pooled, and one of them is rested for a day on a rotation basis. There was also motivation in terms of getting access to sprinklers/drips at subsidy, through linkage with the Andhra Pradesh Micro Irrigation Project (APMIP). The programme has also extended up to 90% support for pipeline network required for water-sharing.

Groups are formed with four to six bore well farmers along with five to eight non-bore well farmers in the neighbourhood to share groundwater and save the crops. The groups are expected to adhere to the following norms:

(i) No farmer should go for drilling of new bore well for 10 years. If it is needed, a decision should be taken in the group after a thorough discussion.
(ii) Saving the rain-fed crops and shifting to growing of drought-tolerant/resistant crops.
(iii) Plantation on farm bunds.
(iv) Digging of one compost pit in one acre of land for improving in situ soil water conservation and soil and moisture conservation (SMC).
(v) Bore well-owning farmers should dig the recharge pit in their fields and thus ensure harvesting of water.
(vi) Farm ponds should be dug in the rain-fed fields. They should ensure storage of rainwater, which could be used to irrigate the crops and reduce the dependence on bore wells.
(vii) Formation of a seed bank.
(viii) Carrying out crop water budgeting (CWB) every year.
(ix) In case there are canals nearby, the farmers should make a representation to the government to avail the water from it.
(x) Dryland farmers should be encouraged to avail micro-irrigation (MI) systems like drip-irrigation scheme benefits through convergence for cultivating vegetable crops.
(xi) Promotion of savings and lending it to members with low interest rate.

Once the groups are formed, representatives of the group are identified. A joint bank account in the name of two leaders will be opened for the purpose of monthly thrift and lending it to members. If the bore well farmers refuse to share water the matter is taken up by the committee. If it is not resolved at this level it can be taken to Mandal Revenue Official (MRO). Irrigation is provided to certain extent of land in both in *Kharif* and *Rabi* seasons depending on the availability of water.

5.2.2 Expected and Perceived Impacts

Based on the institutional modalities, it is expected that groundwater collectivization will be able to provide protective irrigation for selected rain-fed patches in the villages. As a result, it ensures timely sowing, especially during delayed monsoon years, increases cropped area under the pooled bore wells, increases water-use efficiency through the micro-irrigation system, and checks competitive digging of bore wells. These expected impacts or benefits need to be verified with the ground realities. The average annual rainfall in these villages is about 350 mm, making it hard to have any excess water (beyond bare minimum to protect a crop) to share. Besides, these villages faced three consecutive droughts after 2013–2014, i.e. 2014–2015, 2015–2016, and 2016–2017. Of the four villages, Korrakodu initiated the water sharing in 2012, but water sharing was possible only during 2013–2014, as 90% of the bore wells dried up during the later years. This makes the impact assessment difficult in a systematic manner. Here, farmers experiences with water sharing in these villages are presented.

Korrakodu There are nearly 390 bore wells in the village. Though around 1000 bore wells were dug, water is available in 390 bore wells only. After investing on drilling bore wells, a majority of villagers have resolved not to waste any money for digging of wells. According to Mr. Vannurappa: "I owned 2.5 acres of dry land. I am in severe debts due to lack of rains and prevailing drought conditions for the past 10 years. There are no agricultural based labor works available in the village. Due to prevalence of these conditions some farmers seasonally migrate to other areas in search of wage labour. With a hope of getting water I had gone for digging of irrigation bore well. But it disappointed me, as no water was found even after spending Rs. 20,000. At the instance of AF-EC and with the hope of getting water for the crops I joined the water sharing group. Pipelines were laid in our land. They taught us new methods/practices of cultivating crops in dry-lands such as inter-cropping. They provided seeds for the same. I cultivated groundnut, jowar, redgram and vegetables. But there was a long and continuous dry spell and moisture stress situations when the plants were at the flowering stage. At this juncture my neighbouring farmer, Mr. Kuntenna, provided me water for my crops through drip pipes. As a result, I could save my crop from dying and get an extra yield of 40 kg groundnut. I felt very happy about it. I developed the confidence that through this collective sharing/management of water I can protect my rain-fed crop. This is possible as the bore well-owning farmers extend critical and lifesaving irrigation support to those who do not have any irrigation facility. As this is extremely beneficial to the farmers who depend on rains for irrigating their crops, farmers in other villages having similar long dry spell conditions should emulate this system".

M M Halli Of the 50 acres under the group, 12 acres were kept as permanent fallow prior to the formation of the group. Crops like red gram, castor, and groundnut were cultivated in the remaining land. But the crops used to dry due to deficient rainfall. Even the bore well farmers were unable to lay pipelines on account of financial constraints. They used to use sprinkles and damaged pipes borrowed from fellow farmers for irrigating their crops. Groundwater sharing group was formed in 2016, and all the 50 acres of land are getting irrigation from different sources including seepage water from Jeedipalli dam. There is a big change visible among the villagers. With the water management group, protective irrigation is possible now. Farmers are cultivating different crops and getting decent returns. Because of this their life style has changed.

According to Mr. Naganna who owns 5 acres of land with no source of irrigation: "Fetching better yields depends on good rains. If there are good rains the yields from groundnut crop are 4–5 bags per acre. But drought has been prevailing for the past 15 years. I incur a loss of Rs. 6, 500 per year by cultivating groundnut crop. The groundwater sharing group was formed and an agreement was reached between the bore well owners and those who are not having on sharing of water for the crops. I paid Rs. 12, 500 as share capital for my five acres. I cultivated groundnut during Kharif 2016 through pipeline water. My neighbour (farmer) provided me with three irrigations that helped fetch good yield. The yield from five acres was 36 bags, which is more than what I usually get from this land under rain-fed conditions. I got a net income of Rs. 69, 200. I am very happy now".

Sanapa Out of the total 45 acres falling under *Thellagutta* groundwater management group (GMG), 20 acres were kept fallow. In the remaining land redgram, castor, groundnut and millets were cultivated. But due to poor rains, all these crops used to fail. Laying of pipelines was a problem faced by the bore well farmers. They were unable to protect the crops in the rain-fed fields. The key change after the group formation was all the fallow lands were converted into cultivable lands. Crops in dry lands got water security. Instead of waiting for rains, farmers now are growing crops by availing water from the bore wells through pipelines. They are sowing seeds in time. They have shifted to dry land crops which require less amount of water and rainfall to sustain. The bore well farmers are also cultivating red gram, castor, vegetables, and groundnut with good returns. This arrangement gave a hope to dryland farmers. During the first year of the group formation (2015–2016), there was no need for protective irrigation due to good rains. But there was a dry spell in the following *Kharif* (2016–2017), and there was no water in the bore wells to share as they were also drying up.

Mr. Balaraju is a well owner (two wells) with seven acres of land, five acres of wet land, and two acres of dry land. He is also the leader of the water management group. Prior to the group initiative, he used to cultivate paddy (0.75 acre), tomato (3.25 acre), chillis, and groundnut under irrigated conditions. Groundnut was also cultivated under rain-fed conditions. According to him: "I have realized that I should not cultivate paddy and groundnut. I am now cultivating less water intensive and irrigated dry crops like bottle guard, redgram and castor with good returns. I cultivated redgram during 2015–2016 Kharif season in four acres. I got a net income of Rs. 1.80 lakh from it. I cultivated castor in Rabi and got a net income of Rs. 40,000. In the next year i.e., during 2016–2017 Kharif season I cultivated redgram in four acres and got a net income of Rs. 2 lakh".

Kalagalla The 59 acres under *Thurathakonda* water management group were kept fallow in most years prior to the initiative. In the remaining rain-fed lands, red gram, castor, and groundnut are cultivated. A number of watershed activities were undertaken in the village but of little use due to deficient rains. Investments in drilling new bore wells have resulted in farmers ending up in debts on account of depleted groundwater situation. Thirteen farmers (10 ST + 3 OC) having 59 acres of land laid pipelines with the support of agriculture department under RRA project. Of these 13 5 farmers own bore wells and 8 dry land farmers. All the dry land farmers got sprinklers.

In the first year of initiation of the project, 26 acres of 11 rain-fed farmers benefitted through critical irrigation through water sharing. The major crops grown under rain-fed conditions with critical irrigation are groundnut with an intercrop of red gram and castor or castor with red gram. These crops are mostly cultivated during *Kharif* (June). Despite poor rains during 2016–2017, the group could access water from the nearby canal flowing 1.5 km away from their lands. They bought 100 pipes and pumped the water from the canal. Since then farmers, including the dry land farmers, started using the water for irrigation. All this helped to cultivate a variety of crops like vegetables, groundnut, horticultural crops, red gram, castor, horse gram, and jowar and get more returns.

Thus, the experience of bore well collectivization brings out that the initiative has the potential to improve access to water and sustain crops in water stress conditions. These institutions could be game changers in improving equity in groundwater access. However, the major constraint is the harsh natural environment in the sample villages. The meagre annual rainfall of 350 mm may not support groundwater sharing on continuous basis. While natural potential for recharge is limited, external sources of groundwater recharge need to be explored. In this context, there is need for policy interventions that could enhance the access to surface water resources, i.e. through canal distribution systems in these villages. While these aspects are dealt with in detail in the next chapter, the following section assesses some of the older groundwater collectives in semi-arid villages of Anantapuramu district.

5.3   Performance of Groundwater Collectives in Semi-Arid Conditions[3]

In this section we assess two earlier initiatives of groundwater collectives in the semi-arid region of Anantapuramu district. In fact, the institutional modalities of AF-EC interventions are drawn from these initiatives. These interventions were implemented by two different NGOs and supported by different agencies, though the modalities are similar. The assessment is based on two villages, one each representing the two models (Table 5.2). These initiatives are known as social regulation in groundwater management (SRGWM) and Andhra Pradesh Drought Adaptation Initiative (APDAI). These interventions fall in two different semi-arid Mandals with an average annual rainfall ranging between of 545 and 594 mm, i.e. they receive about 200 mm more rainfall when compared to AF-EC villages. These models due to their long-standing nature provide scope for a systematic assessment. Two sample villages (one from each initiative) were selected after visiting a number of villages in the operational areas of these two models. Emphasis was given to select villages with successful implementation so as to assess the operational modalities and the impact of these institutions on access, equity, and sustainability of groundwater use, when implemented under best conditions. In other words, the selection was purposive to remove poor implementation as a factor, so that the possible options for community-based groundwater management (CBGWM) context can be explored. The selection of villages was informed by observations and qualitative discussions made during the preliminary visits. The opinions of the implementing agencies on appropriate fieldwork sites were also sought. Field research was conducted during February and March 2011.

The sample villages vary in size and socio-economic composition (Table 5.3). The geographical area of the villages ranges between 300 and 1900 hectares. Socially, one of the villages is dominated by the other caste (OC) households, while the other has higher proportion of backward caste (BC) households. One of the sample villages has 25% of the households belonging to the scheduled caste/tribe (SC/ST) households. In terms of economic composition, the sample villages are dominated by marginal and small farmers (Table 5.3). Only Madirepalli village has about 50% of the households from medium and large farmers. These variations help in understanding the dynamics of groundwater collectivization in varying socio-economic contexts. About 30 representative sample households were selected from each sample village.

The extent and nature of access to groundwater in the sample villages would highlight the differences in the functioning and performance of the institutions. Both the sample villages depend on groundwater irrigation. The extent of irrigated area ranges between 15% in Gorantlavaripalle to 34% in Madirepalli (Table 5.4). On the other hand, more than 65% of the households in the sample villages have access to wells. Variations in the extent of irrigation could be due to the groundwater

---

[3] This section draws from Reddy et al. 2014.

Table 5.2 Groundwater collectives in semi-arid regions of Anantapuramu district

Village	Mandal	Average rainfall (mm)	Groundwater project	Implementing agency (NGO)	Year of project Initiation	Stage of the project
Madirepalli	Singanamala	545	CWS/SRGWM	RIDS/CWS	2003–2004	Completed
Gorantlavaripalle	Nallacheruvu	594	WASSAN/APDAI	WASSAN/MMS	2007–2008	Completed

Source: Field study using PRA)/FGD methods

Note: *CWS* Centre for World Solidarity, *SRGWM* Social Regulation in Groundwater Management, *APDAI* Andhra Pradesh Drought Adaptation Initiative, *WASSAN* Watershed Support Services and Activity Network, *RIDS* Rural Integrated Development Society, *MMS* Mandal Mahila Samakhya

Table 5.3 Socio-economic composition of the households in the sample villages

Village/farm size	Social categories / No. of households			Total HHs	Sample HHs
	SC/ST	BC	OC		
Madirepalli					
Landless	2	6	1	9	0 (0)
Marginal farmers	26	8	9	43	5 (14)
Small farmers	3	4	30	37	8 (22)
Medium farmers	0	25	30	55	12 (22)
Large farmers	0	7	22	29	5 (17)
Total	**31 (18)**	**50 (29)**	**92 (53)**	**173**	**30 (17)**
Gorantlavaripalle					
Landless	7	0	0	7	0 (0)
Marginal farmers	19	20	5	44	8 (18)
Small farmers	2	30	20	52	19 (37)
Medium farmers	0	0	10	10	3 (30)
Large farmers	0	0	0	0	0 (0)
Total	**28 (25)**	**50 (44)**	**35 (31)**	**113**	**30 (27)**

Source: Field study using PRA/FGD methods
Note: *SC* scheduled caste, *ST* scheduled tribe, *BC* backward caste, *OC* other caste, *HHs* households
Values within simple brackets indicate the % of sample farmers of HHs taken for the study. Values within square brackets indicate the respective percentages of the social groups

Table 5.4 Household access to groundwater in the sample villages

Particulars	Madirepalli	Gorantlavaripalle
No. of households (HHs)	173	113
Average household size	4.2	4.3
Total geographical area (in ha)	307	1064
Area under irrigation (%)	34	15
% of HHs with own wells	43	26
% of HHs sharing wells	45	40
% of HHs with access to wells	88	66
Main occupation	Cultivation	Cultivation

Source: Field survey

potential in the respective villages. It is observed that the SC/ST farmers and marginal and small farmers seem to depend more on sharing water in Gorantlavaripalle, while a large proportion of the OC farmers and large farmers depend on sharing in Madirepalle (Tables 5.5 and 5.6).

Qualitative as well as quantitative research methods have been used for assessing the impacts. Primarily, focus group discussions (FGDs) and household questionnaires were used to elicit the required information. For the purpose of quantitative household data, a detailed questionnaire was used that covered socio-economic, demographic, agriculture, and groundwater management-related issues. In each

Table 5.5 Details of well status of groundwater farmers across social categories in sample villages

Village/caste category	Own well	Water sharing	All
Scheduled caste/tribe	3 (1)	14 (4)	17 (5)
Backward caste	18 (3)	26 (5)	44 (8)
Other caste	53 (11)	38 (8)	91 (19)
Madirepalli total	74 (15)	78 (16)	152 (32)
Scheduled caste/tribe	2 (1)	5 (2)	7 (3)
Backward caste	15 (4)	27 (7)	42 (11)
Other castes	40 (12)	10 (4)	50 (16)
Gorantlavaripalle total	57 (17)	42 (13)	99 (30)

Source: Field study using PRA/FGD methods
Note: Figures in brackets indicate the number of sample groundwater farmer HHs have taken for the study

Table 5.6 Distribution of the sample HHs across farm size and well ownership status (%)

Village	Status of groundwater user Wells	Economic CLASS			
		MF	SF	LMF	Overall
Madirepalli	Owned	7	40	53	48
	Water sharing	31	13	56	52
	Total	19	26	55	100
Gorantlavaripalle	Owned	6	76	18	57
	Water sharing	54	46	0	43
	Total	27	63	10	100

Source: Field survey
Note: *MF* marginal farmers, *SF* small farmers, *LMF* large and medium farmers

village 30 households representing the socio-economic categories of the community were selected. The sample was purposively selected representing well owners and those sharing wells. The sample is representative as the number of sample farmers is in proportion to the actual number of well-owning and well-sharing households.

5.3.1 Institutional Modalities

Both the interventions have been initiated in the semi-arid Mandal of the district, where the extent of groundwater development is quite high. While SRGWM was funded by an external agency, i.e. Aide à l'Enfance de l'Inde (AEI, Luxembourg), the other (APDAI) was supported by the state government (Table 5.7). These models focus on influencing communities through generation of information on groundwater though the degree of using scientific methods used varies.

Table 5.7 Salient features of the community-based groundwater management models

Features	SRGWM	APDAI
Initiative (funding)	External (AEI, Luxembourg)	State government (DoRD)
Implementation	NGOs (CWS + partners)	Govt. + NGO (WASSAN+partners) (*Mahila Samakhya*)
Years of existence by 2011	7	3
Groundwater situation	Scarce	Scarce
Project scale	Small (4 villages)	Small (6 villages)
Key features	Informal regulation	Formal regulation
Scale of operation	Vicinity of wells (within a village)	Area based on the wells (within a village)
Institutional approach	Regulating community through awareness and incentives	Regulating community through semi-scientific information-based awareness and incentives
Operational modalities	Small groups of well owners and dry land farmers followed an intensive approach	Larger group of well owners and dry land farmers covering specific location focus on incentives
Farmers contribution	20% towards micro-irrigation	75%

Source: Field study using PRA/FGD methods and information from the implementing agencies
Note: *DoRD* Department of Rural Development

5.3.2   Social Regulation in Groundwater Management (SRGWM)

SRGWM is an action-research project initiated in 2004 by CWS. The project aims to promote local regulation and management of groundwater resources with equitable access to all families in the communities. The project is expected to develop models to equip the community with drought mitigation preparedness strategies through better water management and regulations at the community level and to support community-based organizations (CBOs) and *Panchayat Raj* institutions (PRIs) in prioritizing the needs of the community for drinking water, irrigation, and other uses, based on the principles of equity.

The project interventions began with a participatory assessment of water resources in the project villages. Growth of groundwater-based irrigation and trends in the groundwater levels were thoroughly discussed and analysed at community meetings, wherein women and men from all households participated. A series of such meetings and interactions helped to better understand frequent failure of bore wells and increasing debts of farmers. For instance, in Madirepalli village, 3 neighbouring farmers dug 13 bore wells in an area of 0.5 acre over a period of 4 years in competition to tap groundwater. The project realized that there is need for changing the mind-set of the farmers from competition to cooperation and to increase the "water literacy" among the farmers for efficient use of water, i.e. reducing non-beneficial and nonrecoverable losses (Perry 2007).

A number of training programmes, exposure visits, and awareness-raising meetings were organized by the grassroots partner NGOs supported by CWS in the project village. Further public awareness and education were carried out through posters, pamphlets, and wall writings. Monitoring of rainfall and groundwater levels in selected bore wells was done regularly by community volunteers using simple manual rain gauge stations installed in the villages and recording the static water levels in ten sample bore wells using an electronic water-level indicator. These data were displayed on a village notice board and discussed in the meetings. It took 3 years (from a total of 7 years) of intensive grassroots work and facilitation to make the community realize the ill-effects of indiscriminate drilling of bore wells and use of groundwater. This helped the community to evolve and agree on the following principles for social regulations and interventions in the village:

- No new bore wells to be drilled in the village
- Equitable access to groundwater for all the families through sharing of wells
- Increasing the groundwater resources by conservation and recharge
- Efficient use of irrigation water through demand-side management

Small groups of farmers were formed between a bore well owner and a group of two to three neighbouring farmers, who did not own bore wells. The bore well owners were motivated to share water, as drilling of new wells in the vicinity of their wells may render theirs dry. Instead, sharing water from his well helps his neighbours while securing his access to water, i.e. "win-win" situation. Farmers were encouraged to practice SGWM by sharing and conserving the resource through micro-irrigation.

5.3.3 Andhra Pradesh Drought Adaptation Initiative (APDAI)

APDAI pilot project is being implemented by the society for elimination of rural poverty (SERP) in collaboration with district collectors and the department of rural development (DoRD). WASSAN is the leading technical agency for this pilot. The institutional modalities of APDAI are similar to that of AF-EC initiative. The project aims at building a case for enabling policy support and investments on critical/protective irrigation and water sharing, focusing on rain-fed farmers. The envisaged model included sharing, social regulation, and controlling competitive digging of bore wells. Further, farmers were provided with pipeline networks for transportation of water to rain-fed farms. Micro-irrigation is promoted to maximize the groundwater use efficiency. Area-based approach involves organizing farmers under common interest groups (CIGs) for a rain-fed patch. In each patch, well owners were convinced of the efficacy of sharing their water with the neighbouring farmers.

5.3.3.1   Achievements of the Initiatives: A Comparative Assessment

Impact assessment is carried out using the information collected at the household level for well owners and water-sharing farmers across farm sizes. Three indicators, viz. access to irrigation, access to critical irrigation, and shifting to less water-intensive crops, were assessed. Besides, awareness and perceptions of the farmers regarding the role and effectiveness of the institutions were also assessed. It may be noted that the sample households include only those farmers having wells or those sharing water from well owners, and hence the proportion of area irrigated is on the higher side when compared to the overall sample villages (Table 5.8). Access to irrigation has gone up in all the sample villages due to sharing and also practising less water-intensive cropping pattern. Across the size classes, increased access to irrigation is greater among marginal and small farmers in the sample villages due to sharing of wells (Table 5.9).

The number of functional wells and households sharing water has gone up in both the sample villages (Table 5.10). This could be due to the better rainfall conditions after 2004 when compared to severe drought conditions (three successive droughts) between 2001 and 2004. Most of the dug wells dried up during 2001–2004, and a few of them revived after 2004. More importantly, investments in new wells are almost absent in the sample villages due to the regulation.

Improved groundwater conditions in the sample villages are also evident from the availability of irrigation in critical periods of crops. Number of farmers reporting availability of groundwater during critical periods has gone up in all the sample villages (Table 5.11). Though this is limited to well owners in Madirepalle, even the

Table 5.8 Changes in percentage area under well irrigation by well status

	Madirepalli (SRGWM)			Gorantlavaripalle (APDAI)		
Status	O	WS	All	O	WS	All
Before	49	8	31	77	18	59
Present	60	63	62	92	64	83
% change	22	688	100	19	255	41

Source: Field survey
Note: Though there was the practice of sharing wells before 2004, there was no area covered as the groups became defunct, consequent to the drying up of wells. Hence, the changes are due not only to increased well sharing but also to the revival of bore wells
O well-owning households, *WS* well-sharing households

Table 5.9 Changes in percentage area under well irrigation by farm size

	Madirepalli (SRGWM)			Gorantlavaripalle (APDAI)		
Status	MF	SF	LMF	MF	SF	LMF
Before	25	39	30	8	66	80
Present	70	58	59	69	85	90
% change	180	49	97	763	29	13

Source: Field survey
Note: *MF* marginal farmers, *SF* small-scale farmers, *LMF* large- and medium-scale farmers

Table 5.10 Changes in access to wells and access to water

Village	Total no. of HHs (population)	Period	Water-sharing HHs	Area under paddy (acres)	Area under irrigation (acres)	Dug wells No.	Area (acres)	Bore wells No.	Area (acres)
Madirepalli	173 (725)	B	8	180	254	59 (2)	4	75 (53)	200 (79)
		A	78	50	491	59 (4)	16	79 (69)	390 (79)
Gorantlavaripalle	113 (487)	B	10	128	140	34 (0)	0	82 (40)	90 (64)
		A	42	80	188	34 (0)	0	84 (46)	138 (73)

Source: Field study using PRA/FGD methods
Note: Figures in brackets are functional wells and % of area. *B* before, *A* after

Table 5.11 Availability of irrigation during critical periods of crop growth by well status (% of farmers)

Status	Madirepalli (SRGWM) O	WS	All	Gorantlavaripalle (APDAI) O	WS	All
Before	0	0	0	60	28	51
Present	10	0	5	77	67	74
% change	–	–	–	29	140	46

Source: Field survey
Note: *O* well owners, *WS* well-sharing households

Table 5.12 Availability of irrigation during critical periods of crop growth by farm size (% of farmers)

Status	Madirepalli (SRGWM) MF	SF	LMF	Gorantlavaripalle (APDAI) MF	SF	LMF
Before	0	0	0	0	56	67
Present	17	75	50	50	79	67
% Change	–	–	–	–	41	0

Source: Field survey
Note: *MF* marginal farmers, *SF* small-scale farmers, *LMF* large- and medium-scale farmers

well-sharing farmers have reported receiving critical irrigation in Gorantlavaripalle (APDAI). Marginal and small farmers are the main beneficiaries in terms of receiving critical irrigation (Table 5.12). This indicates that groundwater institutions have improved the source sustainability and helped in protecting the crops to a large extent. This was possible due to the reduction in the area under paddy (Table 5.13). The reduction in area under paddy is more among large farmers (Table 5.14). The better management of groundwater observed in the case of SRGWM (Madirepalli) is due to the smallness of the group coupled with intensive efforts towards collective strategies when compared to APDAI (Gorantlavaripalle) village.

Table 5.13 Changes in area under paddy by well status (% area)

Crops/status	Madirepalli (SRGWM)			Gorantlavaripalle (APDAI)		
	O	WS	All	O	WS	All
Before						
No crop	13	61	34	13	29	18
Paddy	63	4	38	23	4	17
Groundnut	24	35	29	29	46	34
After						
No crop	5	2	4	3	0	2
Paddy	17	4	12	21	4	16
Groundnut	65	94	78	37	71	48
% change						
No crop	−63	−97	−89	−75	−100	−88
Paddy	−73	0	−69	−7	0	−7
Groundnut	173	171	172	28	54	39

Source: Field survey
Note: *O* well owners, *WS* well-sharing households

Table 5.14 Changes in area under paddy by farm size (% area)

Status/crops	Madirepalli (SRGWM)			Gorantlavaripalle (APDAI)		
	MF	SF	LMF	MF	SF	LMF
Before						
No crop	50	19	36	38	15	10
Paddy	15	55	36	0	16	40
Groundnut	35	26	28	62	31	20
After						
No crop	5	10	0	0	3	0
Paddy	10	16	10	0	16	30
Groundnut	75	68	84 ·	92	40	40
% change						
No crop	−90	−50	−100	−100	−80	−100
Paddy	−33	−71	−73	0	0	−25
Groundnut	114	162	200	50	29	100

Source: Field survey; Note: *MF* marginal farmers, *SF* small-scale farmers, *LMF* large- and medium-scale farmers

The perceptions of the farmers in the sample villages indicate high awareness about the institutions (Table 5.15). However, in both the villages, most of the sample farmers participate in the farmer schools (Table 5.15). Participation rates range between 73% and 100% among sample villages. On the other hand, participation in CWB is low at 40% in the APDAI village (Gorantlavaripalle). But, all the farmers who participated in CWB exercise followed the recommendations in both the villages.

The main benefits derived in the sample villages are awareness about groundwater followed by crop and irrigation methods (Table 5.15). Among the reasons for

Table 5.15 Farmers perceptions on community-based groundwater management (% of farmers)

Perception	Details of perceptions	Madirepalli (SRGWM)			Gorantlavaripalle (APDAI)		
		O	WS	All	O	WS	All
Awareness on groundwater management practices	Awareness	100	100	100	100	100	100
Membership	Yes	93	94	94	76	92	83
Participated in FFS	Yes	100	100	100	82	62	73
Benefits derived	Awareness on crops	100	100	100	82	77	80
	Groundwater methods	100	100	100	82	69	77
	Groundwater awareness	100	100	100	94	77	87
	All of the above	100	100	100	86	74	81
Reasons for not participating (% of not participating farmers)	No tangible benefit	33	38	35	59	85	70
	Not feasible	7	6	6	41	62	50
	Personal reasons	0	0	0	12	15	0
Participated in CWB	Yes	100	100	100	65	8	40
Followed recommendations	Yes	100	100	100	65	8	40
Benefits from groundwater management	Yes	100	100	100	100	100	100
	Conduct of FWS/CWB	100	100	100	100	100	100
	Management of groundwater	100	100	100	82	85	83
	All of the above	100	100	100	94	95	94
Reasons for lack of benefits	Institutions play only advisory role	100	94	97	18	77	43
	Farmers not followed GMC's suggestions	0	6	3	82	23	57

Source: Field survey
Note: *O* well owners, *WS* well-sharing households, *GMC's* groundwater management committees, *FFS* farmer field schools, *CWB* crop water budgeting

non-participation is the absence of tangible benefits followed by those who say that it is difficult to follow or adopt (non-feasibility). While 70% of the non-participating farmers felt that there are no tangible benefits in APDAI village, only 35% of the farmers perceived this reason in the case of SRGWM village (Madirepalli). This perception is greater among the well-sharing farmers when compared to the well owners.

Overall, the performance in terms of physical indicators and farmers' perceptions appears to be better in the case of Madirepalli village (SRGWM). The better performance of SRGWM could be due to the intensive approach it has adopted in promoting water sharing – it has taken almost 3 years to organize the farmers and build awareness before initiating the well-sharing process. Besides, the SRGWM worked with small groups of well-owning and well-sharing farmers, whereas the groups were bigger in the area-based approach of the APDAI. More importantly, the focus was more on well owners as opposed to the entire farming community (the

Table 5.16 Features of the two institutional models

Features	SRGWM	APDAI
Awareness on groundwater situation	High	High
Participation in management	Well owners as well as well-sharing farmers (high)	All the farmers in the well network area (high)
Rules and regulations	Yes (formal)	Yes (formal and binding)
Extent of well sharing	High	High
Cost-sharing	Yes	Yes
Practicing recommendations	High	Low
Additional infrastructure support	Yes (micro- irrigation)	Yes (pipelines and micro-irrigation)
Key to success	Leadership and incentives (subsidy for micro-irrigation)	Incentives (subsidy for pipelines and micro- irrigation)
Impacts on access to water	High	Moderate
Impacts on cropping pattern	High	High
Nature of key impact	Conservation of water and sharing of water	Conservation and sharing of water
Impact on equity	Yes	Yes
Scalability	Poor	Moderate
Sustainability	?	?

Source: Field study (PRA/FGD methods) and reports

majority of whom are prospective well owners) in the case of APDAI village (Table 5.16). This could be an important reason for the poor performance of APDAI when compared to the SRGWM initiative.

5.3.3.2 Lessons for Upscaling

The two models, viz. the SRGWM and APDAI, have adopted social regulation to manage groundwater in the semi-arid region of the district. Social regulations appear to be effective in terms of stopping new bore wells as well as a larger number of households, especially the marginal and small, benefiting from sharing water with well owners. This not only helped in increasing the cropped area but also provided protective irrigation to a number of plots during critical periods, thus saving the crops. This also resulted in equity in the distribution of water and overall welfare improvement. However, there are differences between the two models of social regulation in terms of their effectiveness: the SRGWM appears to be more inclusive and effective when compared to APDAI. One reason could be that the SRGWM is older, followed an intensive approach, and worked with smaller groups of farmers compared to the APDAI. Though APDAI mostly follows the SRGWM approach, it has adopted a broader (area-based) and formal approach involving the department.

APDAI focuses more (though not exclusively) on well owners. This, coupled with the difficulties in organizing larger groups of farmers, has resulted in relatively less effectiveness of the initiative.

Despite the formal approach, participation and rule following are limited in the APDAI. People indicated that there are no tangible benefits from the initiative, and half the farmers felt that the institutional arrangements are not feasible. This view is more conspicuous among those sharing wells. This sceptical nature could be due to scarce groundwater coupled with limited precipitation and replenishing mechanisms. In such uncertain conditions, the larger contribution (75%) from the farmers, which is substantial (total costs are Rs. 8000 to 10,000 per acre, i.e. UD\$135 to 165 per acre), is discouraging farmers.[4] It is observed that the formal process of entering an agreement with the witness of the revenue official has also discouraged some villages from joining the initiative.

The formal approach of APDAI appears good on paper, as it follows an integrated approach of drought adaptation. The integration also involves various departments such as rural development, groundwater, agriculture, etc.; but the feasibility of such integration is doubtful. The approach involves the existing institutions such as the *Mahila Samakhyas*, which provide the assurance of sustenance in the medium run at least. However, at the same time, there is also a danger of acquiring the stamp of a government programme where people look for freebies rather than regulation and contribution.

Sustainability of these initiatives is also a major concern, as they don't have a well-defined exit protocol, while the APDAI appears to be well placed in this regard as its process involves a number of departments and formal institutions. At the same time, it requires strong leadership at the village level to implement and take the initiative forward, especially in the context of people's contribution. In the case of SRGWM, its present success is mainly due to the commitment of NGO partners in the absence of any contribution from the farmers. Besides, in the absence of contribution, the financial sustainability of the initiatives would be a big concern, especially once the external funding stops.

More importantly, the political economy factors come to the fore as these initiatives expand. While the legislation of making groundwater a common property is good on paper, enforcing it at the village level is a major political challenge. Social regulation is a difficult proposition in politically divided communities. It is difficult to presume that large farmers would give up exclusive control on groundwater due to the awareness created. Given that the scientific basis of this awareness is not good enough to protect the farmers from groundwater-related risks, convincing them for adopting SGWM practices is more difficult. Nevertheless, these initiatives represent a small starting point to a "game-changing" groundwater management and may take a longer time span to evolve fully. These initiatives need to work through a number of hurdles, technical, natural, socio-economic, and political.

---

[4]The approach of peoples contribution could provide the much needed ownership and sustainability.

One of the main hurdles is the natural environmental conditions of the region. The benefits from groundwater collectivization are better in the semi-arid villages when compared to arid villages of AF-EC initiatives. At the same time, the uncertainty of rainfall is coupled with limited /no surface water resources proving to be detrimental for scaling up the initiatives even in the semi-arid regions. These constraints are more conspicuous in the context of arid regions with much less annual precipitation. As is the case with groundwater markets, water-sharing institutions will not sustain under severe water constraints. Unless the natural constraints are effectively addressed through artificial groundwater recharge mechanisms, efforts in the direction of groundwater collectivization may have limited success. Therefore, policy interventions that can help to overcome the adverse environmental conditions in the region are a prerequisite or a necessary condition for making groundwater collectivization a reality. Some of the policy options in the context of Anantapuramu district are discussed in the next chapter.

Chapter 6
Making of Climate Smart Communities: Experiences and Lessons

6.1 Introduction

Expecting dramatic improvements or transformation through developmental interventions in agroclimatic context of Ananthapuramu district is an ambitious proposition. Situated in a rain-shadow region with scanty and uncertain rainfall, drought is a near-normal phenomenon with prolonged dry spells. Annual rainfall variation is 30%, and average rainfall ranges between 280 and 750 mm. Probability of timely sowing is 50%, that too with inadequate precipitation (10–40 mm) in most years (Reddy and Reddy 2015). Natural resource base is very fragile with limited availability of surface water resources. Only about 10% of the area is irrigated with 90% dependence of groundwater. And availability of groundwater fluctuates with seasonal rainfall. Of late, climate variability has further aggravated the conditions due to changes in rainfall pattern and distribution. For instance, during 2016 crop season, the district received the lowest rainfall (206 mm), while in September 2017, it experienced heavy downpour receiving equivalent of annual precipitation in just 1 month. Given that more than 80% of the farmers are small and marginal with limited or no access to water, they are increasingly becoming vulnerable and less resilient to face the situation.

Under these conditions AF-EC has been actively pursuing various approaches to minimize, if not mitigate, the agrarian distress in the region over the past three decades. It is one of the first organizations that recognized the importance of WSD to manage the agroclimatic conditions of the region. AF-EC initiated the watershed interventions in the mid-1980s, long before government's watershed programmes started on a large scale. Initially AF-EC focused on conservation of soil, harvesting of rain water, and improving vegetation and biomass. By the late 1990s, it realized that watershed interventions alone are not enough to alleviate poverty and agrarian distress. That is, AF-EC recognized that "though WSD is a necessary condition, it is not sufficient to address the complex agro-ecological dynamics prevailing in the region" and then on AF-EC has started introducing and experimenting

© Springer Nature Switzerland AG 2020
V. R. Reddy et al., *Climate-Drought Resilience in Extreme Environments*,
https://doi.org/10.1007/978-3-030-45889-8_6

watershed-centred developmental initiatives. While livelihoods component has been integral to watershed interventions, AF-EC has been pioneering in making WSD as a drought/climate-resilient strategy. AF-EC was able to take this approach to the community level, where communities are equating WSD with drought resilience.

As evident from our preceding analysis, watershed and related interventions are in the process of transforming the lives of communities in these villages. They have improved water resources, agriculture as a livelihood is stabilizing, and people are able to find alternative livelihoods. Continuation of the process is very likely to strengthen the resource base providing environmental, social, and economic gains. Most importantly farmers have gained in terms of collective or cooperative efforts, decision-making skills, and confidence. AF-EC played a major role in guiding the communities towards a comprehensive and integrated development path that seems to have paid dividends. The gains from such an integrated development path are already evident across the watershed villages. And most of these gains are tangible and significant. Communities acknowledge these gains.

A multipronged approach is being adopted along with the watershed interventions in order to make the communities resilient. The objectives of WSD have been broadened beyond the traditional soil and water conservation and improved vegetation objectives. These include (i) adopting new approaches to RWH and promotion of water-use efficiency, (ii) promoting horticulture and other drought-resilient crop systems and land-use practices, (iii) promoting sustainable agricultural practices, (iv) improved access to credit for poor families (equity) to start off-farm and non-farm IGAs, and (v) capacity building at the community level to promote institution building. Besides, AF-EC is also experimenting with sustainable groundwater management institutions that improve equity in access and protect crops from critical dry spells. Objective-specific activities are designed along with fund allocations from watershed programme funds. AF-EC has been effectively drawing additional resources through integration of other related programme like MGNREGS, Andhra Pradesh Drought Adaptation Programmes (APDAP), etc. While a number of states have been struggling to achieve convergence of programmes, AF-EC could achieve it smoothly through its participatory approaches. In fact, AF-EC has demonstrated to state government the effectiveness of such integration and convergence. Similarly, all the initiatives, be it water related, land related, or livelihoods related, are developed and implemented in an interlinked and integrated manner rather than in isolation. Most of these approaches are not new when looked at them in isolation. The effectiveness of an integrated approach has been demonstrated well. In what follows the approach and experiences from some of the important initiatives that helped building the resilience of the communities are discussed below.

6.2   Integrated Resource Management

6.2.1   Water and Land

Given the low rainfall and arid/semi-arid climate, the central activity of watershed is to harvest and store as much water as possible from rainfall. Traditionally rainwater harvesting (RWH) is taken up under the watershed programme on common lands as on-stream interventions. The most common intervention being check dams. By adopting a ridge to valley approach covering forest lands, other common lands and private lands coupled with net planning at the farmer level, RWH has been moved to farm level with on farm interventions. These include farm ponds, dug out ponds and percolation tanks (PTs). These structures are not huge to meet the normal crop water requirements. Rainwater is collected in farm ponds, stored and being used to irrigate the rain-fed crops during long dry spells between rains. RWH at farm level (farm ponds/dugout ponds) has become an important component in watershed activities. Farm ponds and dug out ponds can be constructed even in small holdings of 1 or 2 acres of land benefitting small farmers. Along with farm ponds and dugout ponds, mini percolation tanks (MPTs) are also promoted. All these interventions help recharge groundwater. While watershed funds support creation of only few such structures, AF-EC could mobilize funds from other programme like NREGS, APDMP, etc., The idea is to create a farm pond on every farmer's field.

The augmented supplies are very limited and hence are fostered with demand management practices. The first in this regard is land-use or cropping pattern changes. Communities are being discouraged from cultivating paddy. Instead, less water-intensive crops like millets, pulses, vegetables, and dryland horticulture are being promoted with necessary support systems. Besides, water-use efficiency is being improved through promotion of micro-irrigation systems (drip and sprinkler). Thus, the strategy of integrating supply (improved water supply) and demand (reduced demand for water) has helped not only expanding the area under protective irrigation but also increased the viability and profitability of farming. Besides, the conjunctive water management practices of combining rainwater with surface or groundwater resources fostered with demand management practices could enhance the water-use efficiency substantially. With these approaches 1–3 lakh acres can be irrigated with 1 thousand million cubic (TMC) of water as against 10,000 acres under flood irrigation (Reddy and Reddy 2015). This has wider applicability across all drought-prone regions in India.

One disadvantage with farm ponds is loss of water due to seepage. That is water may not be available when it is needed most (extreme temperatures). Though farm ponds recharge groundwater, it may not be available at the point of recharge due to all pervasive nature of aquifers. To address this problem farm ponds are now lined with cement so that water is stored for a longer period of time by preventing seepage. This technique has proved to be beneficial in saving the crops during crucial period of plant growth, especially in the case of groundnut – a major crop in this region. For instance, a lined farm pond can retain water for an additional 2–3 weeks.

The yield gains from protective irrigation have been 20–60%. AF-EC has been advocating farm ponds on a large scale, and demonstrations were made to 23,600 farmers during the last 4 years. Consequently, the government has taken up construction of farm ponds on a large scale under MGNREGS. It is the most inexpensive method, and the farmer has much more control over it. Enthused by its success, many farmers are doing it on their own now. This method can be easily replicated in other drought-prone regions also.

Managing water is critical when monsoon is delayed (late sowing) and in the event of prolonged dry spells during the crop season. Technologies and techniques are adopted to overcome this critical water distress. Aqua seed drill and planter + tanker are being promoted to ensure timely sowing when onset of monsoon is delayed. This is an innovation that reduces the risk of losing a crop season, provided there would be enough precipitation during the follow-up period. Similarly, protective irrigation method is being promoted widely to avoid the risk of prolonged dry spells or moisture stress. Studies have shown that more than 50% of droughts occur due to delay/absence of just one rain, i.e. one dry spell of 20–30 days. Prolonged dry spells during crop season cause moisture stress and crops fail. About 80% of the crop failures (droughts) can be saved if two protective irrigations can be given during such dry spells, particularly at the crucial periods of plant growth. Protective irrigation at critical periods also proved beneficial when water is reallocated to less water-intensive crops through area expansion to high-value crops like cotton (Hochman et al. 2017). The efforts of AF-EC have been very effective and spreading fast in the region, especially in the project villages. The efforts have resulted in state government allocating Rs. 1600 million in the budget to scale up protective irrigation measures in 2016–2017 covering the entire state of AP. This could benefit ten million acres of drought-prone land. Government will provide infrastructure, equipment, and support for 2.5 million farmers in AP.

In order to use the available water more efficiently, AF-EC has been researching and developing highly efficient water-saving micro-irrigation (drip and sprinkler) technologies that farmers can use at farm level. AF-EC has developed mobile micro-irrigation technology which can be easily mounted on a tractor or tanker and transported. This equipment can be custom hired to farmers on large scale by the GPs or by women SHGs. The technologies are mobile, farmer friendly, and low-cost. At the same time, cultivation practices are being modified to suit the technologies, i.e. widely spaced crops like red gram and caster crops need just 10,000 litres per acre for one protective irrigation as against 40,000 litres in the case of groundnut. Such low water-intensive mixed crops are being promoted in the place of groundnut and paddy. During 2016–2017, AF-EC demonstrated the practice of protective irrigation with 1820 farmers across 2085 acres of groundnut, red gram, castor, and mango cultivation, after a dry spell of more than 20 days post sowing. Yields increased by 20–40%, and gross returns by Rs. 5000 to Rs. 7500 per acre ata cost of Rs. 600 to 800 per acre. That is, crops can be saved even with lower quantities of water through measured irrigation during dry spells, at crucial periods of plant growth provided appropriate crop systems are adopted.

6.2.2 Sustainable Crop Systems and Practices

Judicious and efficient water management systems would succeed only when land-use or crop practices go in tandem. AF-EC has been promoting dryland horticulture (DLH) on private as well as common lands (block plantation) and fodder development on community lands as part of watershed interventions. Tree species are suitable for local climate, viz. custard apple, mango, goose berry, and tamarind (for fruits). Glyricedia, pongamia (for biomass), neem (for organic pesticides), and ficus varieties like *ravi* (for fodder) are the usual species are drought tolerant and economically beneficial to farmers. Once the saplings grow, they provide assured income and other benefits to the farmer for a longer period of time. They provide green cover, fodder for animals, and increase biomass. DLH is fully supported for 4 years under WSD through provision of saplings and support their upkeep, viz. watering, protecting, etc. WDC employs local villagers to water the block plantations. These crop systems can survive with little protective irrigation, though they have longer gestation to generate income.

Apart from supporting the farmers in nurturing the plantations, multiple mixed crop systems along with DLH were introduced. AF-EC has developed and demonstrated eight drought-resilient rain-fed intercropping systems with millets and pulses, for five consecutive years in about 18,000 acres. They also distributed seeds to farmers covering about 11,620 acres. This has helped reducing the area under groundnut monocrop, which had dropped from 90% to 70% of the cropped area. However, sustaining these initiatives is not easy in the present policy environment, as there is no price support for these crops at the policy level. AF-EC has been lobbying for subsidies for intercrops at par with those extended to groundnut and for inclusion of millets and pulses in fair price shops (FPS), school midday meal programme (MDMP), and integrated child development services (ICDS) and in hostels in government schools and colleges. Replacing rice with locally produced millets and pulses also reduces the food miles in transporting rice.

While changing the policy environment in favour of rain-fed and drought-resilient crop systems is a medium- to long-term strategy, AF-EC has initiated a more sustainable strategy of reducing costs through sustainable agricultural practices. It has been focusing on transforming the conventional high external input destructive agriculture (HEIDA) to low external input sustainable agriculture (LEISA) and non-pesticidal management (NPM). AF-EC intensively campaigns against use of chemicals in agriculture and promotes NPM and various kinds of bio-fertilizers such as vermicomposting, farm composting, liquid fertilizers, mulching, etc. It also promotes crop diversity and crop rotation with the objective of household food and nutritional security on one side and soil health environment and biodiversity on the other hand. Perennial tree crops for fruit, fodder, and biomass as tree crops are more drought tolerant than annual crops and benefit rain-fed farmers as an insurance against droughts. Village-level campaigns on LEISA and NPM are organized in all the 240 project villages under its capacity development programme. Awareness is being created among the farming community on how best they can utilize the locally available bio-ingredients as substitutes to chemical pesticides and fertilizers.

6.3   Institutions and Livelihood Sustainability

Institutions are critical for effective and sustained interventions, especially natural resource-based ones. AF-EC has been promoting and nurturing village-level institutions over the past three decades. Existence of collective efforts and collective institutions have been the main criteria for AF-EC to take up developmental activities in a village. Demonstration of collective or community spirit is taken as demand or need for an intervention at the village level. These collective efforts are evolved into institutions and sustained over the period. The unique features of these institutions are equity, beneficiary contribution, and financial sustainability. All-round equity is ensured by adopting an inclusive approach irrespective of the nature of intervention. For instance, landless and other vulnerable sections are not part of watershed committees (WDC) in most places. In the case of AF-EC, all the socio-economic groups are represented. In fact, priority is given to women and socially weaker groups (SC/ST). While achieving equity is the main concern even in most successful cases (Reddy et al. 2017), it is not a constraint here. There are important lessons in the inclusive approaches adopted by AF-EC. This seems to have come from uncompromising principles and practices adopted at the grassroots level.

Though some of the institutions are programme based, they don't cease to function at the end of the programme. They are either continued, transformed into, or merged with new or existing entities. In the case of watershed institutions, they are either transformed into MACS (NABARD) or merged into Village Organisations (VOs) and then Sasya Mitra Groups: SMGs (IWMP). While watershed guidelines are treated as flexible to address community's requirements, contribution and participation are not compromised. Beneficiary contribution is strictly followed for creating WSD fund for post-programme maintenance. As a result, communities own up the programme and ensure quality of works, which is evident in all the watersheds. Besides, it provides financial sustainability of the institution in the long run and helps maintaining the works. Contributions are collected even during the post-programme phase, when works are carried out on private lands.

Savings and credit activities have become the core activities of these new or transformed institutions. These institutions encourage savings and provide credit to the needy (poor as first priority) to take up livelihood activities. Productive and economically viable livelihood activities are encouraged. Entire village is involved as members and benefits from these initiatives. These institutions have been very successful in creating substantial fund base with high recovery rates in a very short span. Apart from saving and credit, these institutions are responsible for maintaining the watershed works, promoting new and sustainable agricultural practices. These institutions have substantially enhanced the access to credit and reduced the dependency on money lenders, especially for the poor households. MACS and

SMGs/VOs are also venturing into other profitable activities like seed and input supplies, hiring equipment, and other activities, and they operate like banks. They could be transformed into Farmer Producer Organisations (FPOs) without much effort, if at all they are not already operating as FPOs.

Another institutional initiative that could enhance household resilience and ensure equitable access to groundwater resources is the groundwater collectivization institutions that are being tried out on an experimental basis in some villages. Though these initiatives have the potential to enhance equity, they are constrained by the natural conditions, viz. very low rainfall. While these initiatives could be effective in semi-arid (>500 mm rainfall) conditions, the success of these institutions in the arid (<500 mm rainfall conditions) conditions is very limited. While watershed interventions could improve groundwater resources to some extent, they may not be enough to share with rain-fed farmers. Our assessment showed that groundwater sharing is possible mostly during above normal rainfall years. As the above normal rainfall years are becoming rare in the region, there is need for policy interventions in terms of providing surface water for recharging groundwater.

These institutions function democratically with a team of elected (often consensus) office bearers managing the day-to-day affairs. And majority of the members should be women by rule. In fact, there are number of all women committees across the project villages. This is not to say that there are no village politics. But, these politics are treated as given and managed. For instance, in some of the IWMP watershed villages, there was political conflict in electing the watershed chairman; AF-EC worked out a compromise to go without a chairman. VOs or federation of SHGs at the village level are integral to SMGs. More importantly, *Panchayati Raj* Institutions (PRIs) are also integral to these institutions. Due to such integration, the convergence of various government programmes and interventions of various departments as well as agencies has become smooth. This is, in fact, an achievement in convergence. In fact, departments prefer to support these villages and communities, as they are sure of success.

Thus, the comprehensive approach of integrating various dimensions of livelihoods like capacity building, integrated resource management, technologies, and institutions is making communities resilient to drought. Though it may be naive or too early to say that the communities are resilient enough to withstand more than one drought, the process is set and moving in the right direction. Drought resilience has become the priority of the community, and it is very much associated with WSD. It is clear from the experience that the awareness and capacity building at the community level has played an important role in transforming their approach to farming and livelihood strategies. An important aspect is that capacity building is not a one-time activity for a village. A comprehensive approach is adopted, which includes not only watershed related but also alternative livelihoods-related aspects. Institutions are evolved with a long-term perspective, i.e. beyond the programme. Ensuring flexibility in their operations and financial sustainability, i.e. generating resources and managing them as business models rather than as service-oriented models, is helping them to grow. Besides, socio-economic and gender equity are in built, which is unique. And given their organic link with PRIs, which are constitutional bodies, and other CBOs would sustain them in the long run.

Sustainable soil and land management programme (watershed) need to be planned and implemented in a medium- to long-term frame in order to attain their full impact potential. Classical project lifespans of 4–5 years need to be embedded into long-term programme. A combination of capacity and institution building fostered with biophysical and technology investments/promotion has proved more effective. Though this means financial implications at the aggregate level, convergence of different programme could help overcome this. This has been demonstrated very well in the present case. While AF-EC has been focusing on appropriate crop systems, techniques, and technologies, integration of hydrogeology into watershed designing and planning needs more efforts. Our hydrogeology assessment has revealed a mismatch between required and actual RWH interventions. Besides, building awareness on groundwater as a community resource could help making water sharing, which is already being tried, acceptable to the communities. The long-term perspective adopted could not only make communities drought resilient but also make them capable of dealing with climate change impacts in the future, provided they are equipped with more accurate resource knowledge, market, policy support, and climate surveillance information at the village level.

6.4   Lessons and Way Forward

Based on the experiences of AF-EC in making communities drought resilient, we draw some lessons for scaling up and scaling out the initiatives. The idea is to make the interventions more effective. These lessons would be pertinent for all the drought-prone areas in India. It may be noted that some of the aspects raised here fall in the purview of macro policy.

- The experiences of AF-EC clearly brought out that resilience building requires going much beyond watershed guidelines. A number of complementary interventions and support systems need to be integrated, which require additional resources. AF-EC practically demonstrated that through building inclusive institutions and improving the capacities of communities, they could foster convergence of different programmes and resources. They could achieve integration of water management and land-use practices to enhance viability and sustainability of agriculture. Similarly, promotion of farm ponds, micro-irrigation, solar pumps, farm technologies, and sustainable agricultural practices (not part of WSDP) go in tandem in AF-EC project villages, as part of WSD. This is made possible through convergence of WSD with MGNREGS, agricultural programme, etc.
- Inclusive development and equity are at the core of AF-EC initiatives, which is a rare achievement for any agency involved in NRM-based development initiatives. It was observed in the project villages that inclusive (socio-economic and gender) institutions could be created without compromising on broader (entire

village) participation and contribution from the communities. Though achieving equity requires more efforts and time, it ensures effective delivery and sustainability of institutions. The prime role played by women in the project villages helped in achieving overall development (financial, nutritional, social, and human). This is possible only due to a principled stand and commitment of AF-EC. The recent initiatives of groundwater sharing (in non-project villages), when expanded in scale, would further strengthen the inclusive development process.

- Evolving institutions with a long-term vision that are self-sufficient and autonomous rather than creating self-serving programme-specific ones is critical for sustaining NRM initiatives as well as building resilient communities. The experience of creating SMGs and MACS is worth drawing from for scaling up at macro level.
- Apart from the village-level interventions of promoting conjunctive use of water (rain + surface + sub-surface), there is need for broader policy interventions and investments in order to address the extreme distress in these regions. While providing surface irrigation at a large scale may be difficult, filling up the traditional waterbodies in the villages with surface water flows through a canal network or distribution system goes long way in improving the situation.
- For instance, there are above 3000 small and medium waterbodies spread across in Ananthapuramu district. All these waterbodies need to be revived and converted into percolation tanks. These are common property resources of the villages, and all the households have the right to use the water. If these can be filled with *Tungabhadra* or *Hundri-Neeva Sujala Sravanthi* (HNSS) project water, all farmers can use it for protective irrigation through bore wells, as percolation tanks recharge and stabilize groundwater. Communities feel that it can support much needed protective irrigation needs in a most efficient way. This is necessary condition for the evolution and success of groundwater collectivization interventions and institutions in these harsh environmental conditions.
- In the absence of assured water availability to protect at least one crop in a year, it would be difficult to sustain farming in the district. For example, one drought can wipe out all savings and income. In this context, there is a need to create water grid exclusively for Anantapuramu, i.e., "Anantapuramu water grid". It is estimated that at least 25 TMC of water is required to fill all the waterbodies in the district (Reddy 2017). A canal distributary system needs to be developed connecting all the waterbodies and supply water from the surface reservoirs like *Tungabhadra* and *Hundri-Neeva* projects. These canal systems recharge the groundwater on their way to fill the tanks. For this, assessing aquifer geometry across different regions of the district is required in order to understand the recharge potential in each region.
- The importance of groundwater in supporting the livelihoods in the district need not be overemphasized. Strengthening the surface waterbodies and connecting them to the canal systems would increase groundwater recharge and availability. This needs to be used judiciously and equitably to ensure the benefits to all and forever. For this, three important policy measures are required. These include (a)

conjunctive use of surface and groundwater, (b) increase water-use efficiency through micro-irrigation, and (c) equitable distribution through sharing of groundwater. And groundwater collectivization can be instrumental in achieving these objectives.

- Conjunctive use of surface and groundwater needs to be promoted throughout the district by capturing and conserving the in situ moisture and by increasing the crop coverage and other practices like mulching. Encourage rainwater harvesting at the farm level through farm ponds and dugout ponds. There is need for creating farm ponds for every 3–5 acres of crop area (Reddy and Reddy 2017). And these ponds need to be lined in order to reduce seepage. The stored water can be used to protect the crops prior to switching to groundwater. During drought years groundwater should be used only to protect the crops during critical periods. At the same time, water-use efficiency should be enhanced through micro-irrigation, making micro-irrigation mandatory for financial support towards bore wells and farm ponds.

- The most important and difficult aspect of groundwater management is its distribution equitably. Given the skewed property rights regime and the lumpy nature of capital investment required groundwater access is biased in favour of large and medium farmers depriving the small and marginal farmers. While making groundwater as a common property by legislation is an ideal approach, it requires a long-term approach of bringing legal and legislative reforms. In the short to medium term, farmers may be encouraged to share groundwater through some social regulations like restricting the number of bore wells, ban on new bore wells, etc. Such regulations are being tried successfully in some parts of the district by AF-EC itself, and they can be scaled up with the support of the state government and the NGOs. For instance, state government should strengthen canal distribution system by bringing water from irrigation dams and recharge groundwater through filling up surface waterbodies. And rural development interventions or welfare interventions at the village level may be linked to acceptance of such regulations.

- While most of the interventions are bearing fruit in terms of moving towards sustainable cropping systems, DLH, increased yields, etc., these developments when scaled up could create serious marketing issues. As the supply expands, prices may crash leaving farmers high and dry. The marketing and value chain management need to be planned in advance. This needs policy and institutional changes at the higher level. Farmers need to be connected to the markets beyond their vicinity. Where potential for productivity gains exists, market constraints seem to be the main limiting factor for technology penetration. This might be the next issue that AF-EC needs to address with some out of the box initiatives, where farmers can get the benefits from value addition.

- The 2003 Agriculture Produce Marketing Committees (APMC) Act provisions are adopted in a diluted form in most states (Chand 2016). Some new market policies of 100% foreign direct investment (FDI) in domestic trading of processed foods and establishment of e-national agriculture market (e-NAM) were introduced during 2016 (Rao et al. 2017). These reforms are proving to be effective in improving the price realization of farmers in a short span (Chand 2017). Ensuring such reforms reach remote farmers is critical for their sustenance.

- Resilience-building initiatives in the drought-prone regions need to be built around WSD, as it helps strengthening natural resource base (necessary condition). The recent developments of diluting WSDP at the central level may not augur well given the necessity of watershed interventions in these regions. The limited success of watershed interventions at the national level is often used to dilute the programme. But, the reality is that watershed interventions are neither designed scientifically nor implemented comprehensively. The effectiveness of systematic and comprehensive implementation is evident from the experiences of PIAs like AF-EC and Watershed Organisation Trust (WOTR). That is, there is need for strengthening the scientific base and implementation modalities of the initiatives. Taking a comprehensive view by integrating science and socio-economic aspects is necessary for sustainability of the interventions in the long run. Though the success and sustainability of the programme is mainly due to human and social capital built around resource and economic exigencies, science-based awareness pertaining to hydrogeology is limited. This is necessary for ensuring the effectiveness of the initiatives at the national level.
- Another important area of macro policy reform is the effective implementation of minimum support price (MSP) for rain-fed crops like millets, pulses, etc.; AF-EC has been trying to push this along with interventions for raising the demand for these crops (introducing in MDMS (midday meal schemes) and through PDS (public distribution system)) at the state level. Since fixing MSP and its implementation is under the purview of the central government, macro reforms are needed to protect the farmers from drought-prone regions. Nevertheless, state governments can provide some support and relief at least to overcome the distress situations.
- The long-term impacts can be captured better when there is good baseline information. Due importance needs to be given for conducting a robust baseline survey (BLS) in the design of the programme. The baseline needs to be comprehensive, including technical as well as socio-economic aspects of the sample sites. The baseline should also consider the monitoring and evaluation framework of the programme.

References

ACIAR (2015) Project Report on Developing multi-scale adaptation strategies for farming Communities in Cambodia, Lao PDR, Bangladesh and India. Australian centre for International Agricultural Research (ACIAR), Australia

Batchelor CA, Singh K, Rama Mohan Rao CH, Butterworth C (2003) Watershed development: a solution to water shortages or part of the problem? Land Use Water Resour Res 3:1–10

Bhatia B (1992) Lush fields and parched throats, political economy of groundwater in Gujarat. Econ Pol Wkly 27(51–52):19–26

Burke J, Moench M, Sauveplane C (1999) Groundwater and society: problems in variability and points of engagement. In: Salman M, Salman A (eds) Groundwater – legal and policy perspectives, World Bank technical paper, No. 456. The World Bank, Washington, DC, pp 31–52

Carpenter S, Walker B, Anderies JM, Abel N (2001) From metaphor to measurement: resilience of what to what? Ecosystems 4:765–781. https://doi.org/10.1007/s10021-001-0045-9

Chand R (2016) e-Platform for national agriculture market. Econ Pol Wkly 51(28):15

Chand R (2017) Doubling farmers' income: strategy and prospects. Indian J Agric Econ 72(1):1

Dhawan BD (1995) Magnitude of groundwater exploitation. Econ Pol Wkly 30(14):769–775

Gallopín G (2007) Linkages between vulnerability, resilience, and adaptive capacity, workshop Formal approach to vulnerability, Potsdam Institute for Climate Impact Research September 13–14, Potsdam

GoAP (n.d.) Watershed Development (PMKSY–Watersheds): Department of Rural Development

GoAP (2015) 5th Minor Irrigation Census (2013–14), Directorate of Economics Statistics, Andhra Pradesh, Vijayawada

GoI (2008) Common guidelines for watershed development projects. Ministry of Rural Development Department of Land Resources, New Delhi

GoI (2019a) Regional review meeting of Southern Region States on WDC-PMKSY, Department of Land Resources, Ministry of Rural Development, Bengaluru, 29–30 July

GoI (2019b) Department of Land Resources, Ministry of Rural Development

Hochman Z, Horan H, Reddy DR, Sreenivas G, Tallapragada C, Adusumilli R, Gaydon DS, Roth CH (2017) Smallholder farmers managing climate risk in India: 1. adapting to a variable climate. Agric Syst 150:54–66

http://iwmp.ap.gov.in/WebReports/Content/Programmes.html

https://core.ap.gov.in//CMDashBoard/UserInterface/IWMP/IWMPREPORT.aspx

https://dolr.gov.in/

https://www.krishaksarathi.com

https://www.krishaksarathi.com/watershed-development-programme.html

https://www.nabard.org/demo/auth/writereaddata/File/21%20_WATERSHED_MANAGEMENT.pdf

© Springer Nature Switzerland AG 2020   161
V. R. Reddy et al., *Climate-Drought Resilience in Extreme Environments*,
https://doi.org/10.1007/978-3-030-45889-8

IPCC (2007) Climate change 2007: impacts, adaptation and vulnerability. Contribution of Working Group II to the fourth assessment report of the Intergovernmental Panel on Climate Change, Parry ML, Canziani OF, Palutikof JP, van der Linden PJ, Hanson CE (eds), Cambridge University Press, Cambridge, 976pp

Libecap GD (1997) Contracting for property rights. Karl Eller Centre and Department of Economics, University of Arizona/National Bureau of Economic Research, Tucson, Arizona/ Cambridge, MA

Llamas R, Martínez-Santos P (2005) Intensive groundwater use: silent revolution and potential source of social conflicts. J Water Resour Plann Manag 131(5):337–341

LNRMI (2012) Evaluation Study of Watershed projects (Preparatory Phase) in Maharashtra, Project Report, Livelihoods and Natural Resource Management Institute, Hyderabad, India, August

Moench M (1992) Chasing the water table: equity and sustainability in groundwater management. Econ Pol Wkly 27(51–52):A171–A177

Perry C (2007) Efficient irrigation; inefficient communication; flawed recommendations. Irrig Drain 56(4):367–378

Polak P (2004) Water and the other three revolutions needed to end rural poverty. Invited paper in the World Water Week, Stockholm 15–20 August, Preprint, Stockholm International Water Institute, Stockholm

Rao NC, Sutradhar R, Reardon T (2017) Disruptive innovations in food value chains and small farmers in India. Indian J Agric Econ 72(1)

Ravindranath NH, Chaturvedi RK, Joshi NV et al (2011) Mitig Adapt Strat Glob Chang 16(2):211–227

Reddy VR (1999) User valuation of renewable natural resources: user perspective. Econ Pol Wkly 34(23):1435–1444

Reddy VR (2004) Managing water resources in India: a synoptic review. J Soc Econ Dev 6(2):176–193

Reddy VR (2005) Costs of resource depletion externalities: a study of groundwater overexploitation in Andhra Pradesh, India. Environ Dev Econ (Cambridge) 10(Part IV):533

Reddy YVM (2017) Ananta Prasthanam: Magnakarta of a drought District – 2 (Telugu). Kadalika Publishers, Ananthapuramu

Reddy VR, Reddy PP (2005) How participatory is participatory irrigation management: a study of water user associations in Andhra Pradesh. Econ Pol Wkly 40(53):5587–5595

Reddy YVM, Reddy TY (2015) Conjunctive water management: an approach & technologies for successful cropping under rainfed conditions in Anantapur District. Accion Fraterna Ecology Centre, Ananthapuramu

Reddy VR, Reddy MS (2017) Water and sustainable livelihoods: need for an Integrated Watershed Management (jointly). In: Srinivasa Raju K, Vasan A (eds) Sustainable holistic water resources management in a changing climate. Jain Brothers, New Delhi

Reddy VR, Syme G (2015) Integrated assessment of scale impacts of watershed interventions: assessing hydro-geological and bio-physical influences on livelihoods edited with Geoffrey J. Syme, Elsevier INC, October, 2015

Reddy VR, Reddy MG, Reddy YVM, Soussan J (2004) Assessing the Impacts of watershed development programme: a sustainable rural livelihoods framework (Jointly). Indian J Agric Econ 59(3):59–84

Reddy VR, Reddy MG, Soussan J (2010) Political economy of watershed management: policies, institutions, implementation and livelihoods. Rawat Publishers, Jaipur

Reddy VR, Syme G, Ranjan R, Pavelic P, Reddy MS, Rout SK, Sreedhar A (2012) Scale issues in meso- watershed development: farmer's perceptions on designing and implementing the common guidelines. Working paper, No. 2. Livelihoods and Natural Resource Management Institute, Hyderabad

Reddy VR, Reddy MS, Rout SK (2014) Groundwater governance: a tale of three participatory models in Andhra Pradesh, India. Water Altern 7(2):275–297

Reddy VR, Reddy SS, Chiranjeevi T, Rout SK (2017) Long-term effects of German development co-operation in soil improvement and sustainable land & water management in India, draft report, Livelihoods and Natural resource Management institute, Hyderabad

Reddy VR, Cunha DGF, Kurian M (2018a) A water–energy–food nexus perspective on the challenge of eutrophication (jointly). Water 10:101. https://doi.org/10.3390/w10020101

Reddy VR, Palanisamy K, Reddy MS (2018b) Tank rehabilitation in India: review of experiences and strategies (Jointly). Agric Water Manag:209, 32–243. https://doi.org/10.1016/j.agwat.2018.07.013

Shah T (2004) Groundwater and human development: challenges and opportunities in livelihoods and environment. Invited paper in the World Water Week, Stockholm 15–20 August, Preprint, Stockholm International Water Institute, Stockholm

Shah T (2009) Taming the anarchy: groundwater governance in South Asia. The Resources for the Future Press, Washington, DC

Sharma KD (2009) Groundwater management for food security. Curr Sci 96(11):1444–1447

Uitto JI (2019) Sustainable development evaluation: understanding the nexus of natural and human systems. In: Julnes G (ed) Evaluating sustainability: evaluative support for managing processes in the public interest, New directions for evaluation, 162, pp 49–67

Valencia Statement (2004) In Sahuquillo A, Aliaga R, Cortina LM, Sanchez Vila X (eds), Groundwater intensive use, pp 399–400. Selected papers of the SINEX Conference, Valencia, Spain, 10–14 December 2002. A.A. Balkema Publishers, Leiden

Verma S, Krishnan S, Reddy VA, Reddy KR (2012) Andhra Pradesh farmer managed groundwater systems (APFAMGS): a reality check. Water policy research highlight; IWMI-TATA Water Policy Programme

Printed in the United States
by Baker & Taylor Publisher Services